REGULATION OF DAM SAFETY

AN OVERVIEW OF CURRENT PRACTICES WORLDWIDE

REGLEMENTATION DE LA SECURITÉ DES BARRAGES

REVUE INTERNATIONALE DES PRACTIQUES ACTUELLES

ICOLD Bulletin 167, **Regulation of Dam Safety: An Overview of Current Practices Worldwide**, provides a comprehensive review of legal and regulatory arrangements for the safety of dams among the countries represented at ICOLD. As such, this review is essentially a snapshot of the situation being in place at the end of the first and the beginning of the second decade in the 21st century.

This Bulletin is useful not only to these countries which have weak or non-existent legal and regulatory dam safety frameworks but also to these jurisdictions which are considering changes and improvements to existing legislation and regulations.

Le Bulletin 167 de la CIGB, **Reglementation de la Securité des Barrages: Revue Internationale des Practiques Actuelles**, fournit une revue complète des dispositions légales et réglementaires en matière de sécurité des barrages au sein de pays représentés à la CIGB. À ce titre, cette revue est une photographie instantanée de la situation autour des années 2010.

Il peut être utile non seulement pour les pays qui ont un cadre juridique et réglementaire réduit ou inexistant pour la sécurité des barrages, mais également à ceux qui envisagent des changements et des améliorations de leurs législations et réglementations en vigueur.

I0131787

INTERNATIONAL COMMISSION ON LARGE DAMS
COMMISSION INTERNATIONALE DES GRANDS BARRAGES
6 quai Wattier – 78500 Chatou (France)
Telephone : + 33 6 60 53 07 31
http://www.icold-cigb.org.

Cover illustration: World map – participating countries
Couverture: Carte du monde – Pays ayant participé à l'enquête

CRC Press/Balkema is an imprint of the Taylor & Francis Group, an informa business
© 2025 ICOLD/CIGB, Paris, France

Typeset by codeMantra

Published by: CRC Press/Balkema
Schipholweg 107C, 2316 XC Leiden, The Netherlands
e-mail: Pub.NL@taylorandfrancis.com
www.routledge.com – www.taylorandfrancis.com

Original text in English
French translation by Michel Poupart
Layout by Nathalie Schauner

Texte original en anglais
Traduction en français par Michel Poupart
Mise en page par Nathalie Schauner

ISBN: 978-1-032-45605-8 (Pbk)
ISBN: 978-1-003-37779-5 (eBook)

COMMITTEE ON DAM SAFETY

COMITE DE LA SECURITE DES BARRAGES

Chairman/Président

Canada   Przemyslaw A. ZIELINSKI

Members/Membres

Argentina / Argentine	Francisco GIULIANI
Australia / Australie	Shane MCGRATH
Austria / Autriche	Elmar NETZER
Brazil / Brésil	Fabio De Gennaro CASTRO
Bulgaria / Bulgarie	Dimiter TOSHEV
Canada	Des HARTFORD
China / Chine	Zeping XU
Czech Republic / République Tchèque	Jiri POLÁČEK
Finland / Finlande	Eija ISOMAKI
France	Michel POUPART
Germany / Allemagne	Hans-Ulrich SIEBER
Iran	Mohsen GHAEMIAN
Italy / Italie	Carlo RICCIARDI
Japan / Japon	Kentaro KIDO
Korea / Corée (Rép)	SHIN Dong-Hoon
Latvia	Sigita DISLERE
Netherlands / Pays Bas	J.P.F.M. JANSSEN
Norway / Norvège	Grethe Holm MIDTTØMME
Pakistan	Azhar Salim SHEIKH
Portugal	Antonio Da Silva GOMES
Russia / Russie	Evgenyi BELLENDIR
Slovenia / Slovénie	Andrej KRYZANOWSKI
Serbia / Serbie	Ignjat TUCOVIC
Slovakia / Slovaquie	Peter PANENKA
Slovenia / Slovénie	Nina HUMAR
South Africa / Afrique du Sud	Ivor SEGHERS
Spain / Espagne	Juan Carlos DE CEA
Sri Lanka	Badra KAMALADASA
Sweden / Suède	Maria BARTSCH
Switzerland / Suisse	Marc BALISSAT
Turkey / Turquie	Tuncer DINÇERGÖK
United Kingdom / Royaume Uni	Andy HUGHES
United States / États Unis	C. Gus TJOUMAS

SOMMAIRE	CONTENTS

TABLE DES MATIERES

TABLE OF CONTENTS

TABLEAUX & FIGURES

FIGURES

TABLES & FIGURES

FIGURES

1. PRÉAMBULE

Ce rapport fournit une revue complète des dispositions légales et réglementaires en matière de sécurité des barrages au sein de pays représentés à la CIGB. À ce titre, cette revue est une photographie instantanée de la situation autour des années 2010. Ce rapport peut être utile non seulement pour les pays qui ont un cadre juridique et réglementaire réduit ou inexistant pour la sécurité des barrages, mais également à ceux qui envisagent des changements et des améliorations de leurs législations et réglementations en vigueur.

Il est notable de constater que, après de nombreuses années de réflexions sur les avantages et inconvénients des méthodes d'évaluation des risques pour la sécurité des barrages, cette approche commence à faire son apparition dans la réglementation de certains pays. Il est prématuré à ce jour de conclure que cette tendance va se poursuivre et s'étendre à d'autres pays. Le cas échéant, ce sera un développement très important et le Comité de la Sécurité des Barrages a l'intention de suivre la situation et de mettre à jour ce rapport.

Przemyslaw A. Zielinski
Président du comité de la sécurité des barrages

1. FOREWORD

This report provides a comprehensive review of legal and regulatory arrangements for safety of dams among the countries represented at ICOLD. As such, this review is essentially a snapshot of the situation being in place at the end of the first and the beginning of the second decade in the 21st century. The report may be useful not only to these countries which have weak or non-existent legal and regulatory dam safety frameworks but also to these jurisdictions which are considering changes and improvements to existing legislation and regulations.

It is noteworthy to observe that after many years of considering benefits and shortcomings of formal dam safety risk assessment the approach informed by assessment of risks is slowly finding its way into regulation of dam safety in some countries. It is premature at this time to conclude that this trend will continue and spread to other countries. If it does, it will be a very important development and the intention of the CODS is to monitor the situation and issue the update of this report.

Przemyslaw A. Zielinski

Chairman, Committee on Dam Safety

2.   REMERCIEMENTS

Le Comité de la Sécurité des Barrages et la Direction de la CIGB tiennent à saluer la contribution des membres du Groupe de Travail du Comité et le soutien logistique des entreprises ou organismes auxquels ils appartiennent. Le texte final du Rapport est le résultat de l'effort collectif des membres du Groupe :

1. Mr. Jiří Poláček, Responsable du Groupe de Travail, Directeur Senior, Vodni Dila, République tchèque – aides financière et en nature apportée par Vodni Dila;

2. Ms. Maria Bartsch, Responsable Sécurité Barrage, Svenska Kraftnät - aides financière et en nature apportée par Svenska Kraftnät;

3. Mr. Fabio de Gennaro Castro, Vice-président, Comité Brésilien des Barrages - aides financière et en nature apportée par le comité brésilien des barrages;

4. Ms. Grethe Holm Midttømme, Ingénieure en Chef (Sécurité des barrages), The Norwegian Water Resources and Energy Directorate - aides financière et en nature apportée par The Norwegian Water Resources and Energy Directorate;

5. Ms. Nina Humar, Responsable du Comité Scientifique du Comité Slovène des Grands Barrages (SLOCOLD) - aides financière et en nature apportée par SLOCOLD;

6. Ms. Eija Isomäki, Spécialiste en Sécurité des Barrages, ELY Centre for Häme - aides financière et en nature apportée par ELY Centre for Häme;

7. Dr Hans-Ulrich Sieber, Directeur, Administration des Réservoirs de l'État de Saxe, Allemagne - aides financière et en nature apportée par l'Administration des Réservoirs de l'État de Saxe, Allemagne;

Ces remerciements s'adressent également au Dr Andy Hugues pour le temps passé à la mise au point finale et l'aide apportée à l'édition du rapport.

Traduction française par Michel POUPART

2. ACKNOWLEDGMENTS

The Committee on Dam Safety and the ICOLD Executive gratefully acknowledge the contribution of members of the Committee's Working Group and the support provided by their sponsoring organizations. The final text of the Report is the result of the collective effort of the following members of the Group:

1. Mr. Jiří Poláček, Working Group Leader, Senior Manager, Vodni Dila, Czech Republic – financial and in kind assistance provided by Vodni Dila;

2. Ms. Maria Bartsh, Dam Safety Officer, Svenska Kraftnät - financial and in-kind assistance provided by Svenska Kraftnät;

3. Mr. Fabio de Gennaro Castro, Vice-President, Brazilian Committee on Dams - financial and in kind assistance provided by Brazilian Committee on Dams;

4. Ms. Grethe Holm Midttømme, Head Engineer (Dam Safety), The Norwegian Water Resources and Energy Directorate - financial and in kind assistance provided by The Norwegian Water Resources and Energy Directorate;

5. Ms. Nina Humar, Head of Scientific Committee of Slovenian Committee on Large Dams (SLOCOLD) - financial and in-kind assistance provided by SLOCOLD;

6. Ms. Eija Isomäki, Dam Safety Specialist, ELY Centre for Häme - financial and in-kind assistance provided by ELY Centre for Häme;

7. Dr Hans-Ulrich Sieber, Managing Director, State Reservoir Administration of Saxony, Germany - financial and in-kind assistance provided by State Reservoir Administration of Saxony, Germany;

The final acknowledgement is extended to Dr. Andy Hughes who volunteered his time and provided editing support.

3. INTRODUCTION

En 2008, le Comité sur la Sécurité des Barrages (CODS) de la CIGB a constaté qu'il n'existait pas de revue internationale à jour concernant la législation de la sécurité des barrages (lois, règlements, normes, etc.), leur surveillance et leur classification. Un Groupe de Travail a en conséquence été chargé de rédiger un rapport sur ce sujet.

Le rapport de la Banque Mondiale "Les cadres réglementaires pour la sécurité des barrages : une étude comparative" (2002) [1] avait abordé ce sujet sous quatre angles 1) la forme juridique des réglementations, 2) les dispositions institutionnelles pour réglementer la sécurité du barrage, 3) les prérogatives des autorités fixant la réglementation 4) le contenu de la réglementation. Toutefois, il ne couvrait pas en détail des aspects de la sécurité tels que la classification et la surveillance des barrages.

L'étude de la Banque Mondiale a décrit les cadres réglementaires dans 22 pays qui ont été sélectionnés principalement en fonction de la disponibilité de l'information. Elle a d'une part mis en évidence les similitudes et les disparités et d'autre part émis des recommandations sur le contenu attendu d'un cadre réglementaire. Dans la revue effectuée par le Comité de la Sécurité des Barrages de la CIGB le nombre de pays ayant répondu a plus que doublé et dans le même temps les informations recueillies auprès de chaque pays sont plus nombreuses.

Le rapport de la Banque Mondiale a souligné que "la sécurité des barrages est un concept dynamique et évolutif" et devrait être traité en conséquence. De fait, au cours des travaux de la présente étude, il a été constaté que dans plusieurs des pays étudiés des développements importants du cadre juridique de la sécurité du barrage ont eu lieu au cours de la dernière décennie, ou sont en cours. En conséquence une partie des informations du rapport de la Banque Mondiale, datant dans la plupart des cas des années 1990 ou antérieures, n'est plus à jour.

Le Groupe de Travail a été chargé de recueillir et de décrire les grands principes des cadres juridiques pour la sécurité des barrages dans différentes parties du monde. Les systèmes juridiques ainsi que la répartition des responsabilités pour la sécurité et la surveillance des barrages, les modalités de contrôle indépendant (par l'autorité responsable, ou par une entité autre que le propriétaire), la classification et les référentiels techniques ont fait l'objet d'une attention toute particulière.

L'objectif a été de mettre à disposition les informations de base sur les pratiques existantes en matière de gestion de la sécurité des barrages. En terme de résultat, l'étude donne un aperçu global des principales dispositions applicables aux cadres de la sécurité des barrages. Les différences et tendances observées font l'objet de commentaires, et une liste de référence permet d'accéder à des informations plus détaillées sur chaque pays.

3. INTRODUCTION

In 2008 the Committee on Dam Safety (CODS) in ICOLD noted that a current worldwide overview relating to dam safety legislation (acts, regulations, standards etc.), dam supervision and dam classification (categorization) did not exist. A Working Group for Dam Safety Legislation and Dam Classification was established to produce the document.

The World Bank Report "Regulatory Frameworks for Dam Safety: A Comparative Study" (2002) [1] dealt with this matter divided into four issues; 1) the legal form of the regulations, 2) the institutional arrangements for regulating dam safety, 3) the powers of the regulating entity and 4) the contents of the regulatory scheme. However, it did not cover in detail aspects of dam safety such as dam classification and surveillance of dams.

The World Bank study described the regulatory frameworks in 22 countries which were selected mainly on the availability of information. Similarities and differences were highlighted and recommendations on what a regulatory framework for dam safety should contain were given. In the survey presented now by the CODS of ICOLD the number of countries has more than doubled and at the same time the extent of active contributions from each country has been increased.

The World Bank Report emphasized that "dam safety is a dynamic, evolving concept" and should be treated accordingly. In fact, during the work with the present study it has been found that in several of the countries studied significant development of the legal framework of dam safety has taken place during the last decade or is ongoing. As a result, part of the information in the World Bank Report that in most cases date from references from the 1990's or earlier, is no longer up to date.

The CODS Working Group was given the task to investigate and document the main principles of legal frameworks for dam safety in different parts of the world. Special attention was given to legal systems and the distribution of responsibilities for dam safety and surveillance of dams, arrangements for independent supervision of dam safety (by the responsible authority, or by a party/body other than the dam owner), classification of dams and technical frameworks.

The goal has been to make basic information available about existing practices in dam safety management. The study has resulted in an overview of the main arrangements for dam safety frameworks. Some differences and trends are discussed, and references are given to more detailed information on each of the countries in the study.

4. RECUEIL DES DONNÉES

Pour rassembles les informations, un questionnaire simple a été élaboré (annexe A). Les questions portaient sur :

- La législation de la sécurité des barrages (textes réglementaires et leurs références, dispositions organisationnelles et attribution des responsabilités pour la sécurité des barrages, la supervision et la surveillance) ;

- Les référentiels techniques en matière de sécurité des barrages (recommandations etc. avec leurs références, pratiques d'évaluation du niveau de sécurité des barrages) ;

- La classification des barrages (exigences légales, critères de classification et nombre de classes, applicabilité aux différents types de barrages et ouvrages hydrauliques).

En 2009, le questionnaire a été envoyé à une sélection de plus de 60 pays membres actifs de la CIGB de tous les continents. Pour les quelques 30 pays qui étaient membres du comité de la sécurité des barrages à l'époque, le représentant national a été invité à fournir des informations au groupe de travail. Pour les pays n'ayant pas de représentant dans le comité de la sécurité des barrages, les comités nationaux de la CIGB ont été interrogés directement.

Une trentaine de pays ont répondu dès la première année suite à l'envoi du questionnaire, soit environ 50% des pays. Les contributions supplémentaires obtenues suite aux rappels effectués lors des réunions annuelles de la CIGB et aux relances par courrier électronique permettent d'aboutir finalement à 44 pays à l'horizon 2012.

Ces pays se distribuent géographiquement de la façon suivante :

- Europe 25

- Asie 8

- Amérique du Nord 3

- Amérique du Sud 3

- Afrique 3

- Océanie 2

4. DATA COLLECTION

To gather information a simple questionnaire was developed, see Appendix A. The questions were centered on:

- Legislation on dam safety (regulations with references, organizational arrangements and responsibilities for dam safety, supervision and surveillance)

- Technical framework concerning dam safety (guidelines etc. with references, practice for evaluation of dam safety)

- Classification of dams (legal requirements, criteria for classification and no. of classes, applicability to different types of dams and hydraulic structures)

In 2009 the questionnaire was sent to a selection of more than 60 active ICOLD member countries from all continents. For the approximately 30 countries that were members of CODS at the time, the national delegate was asked to provide the information to the working group. For countries without representation in the CODS the national committees of ICOLD were approached.

Within the first year, after the survey was initiated input was received from about 30 countries (50% of the selected countries). Participation in the investigation was promoted at ICOLD annual meetings and via e-mail reminders. This resulted in additional contributions and a total of 44 countries have provided input up to 2012.

The participating countries are distributed geographically as follows:

- Europe 25

- Asia 8

- North America 3

- South America 3

- Africa 3

- Oceania 2

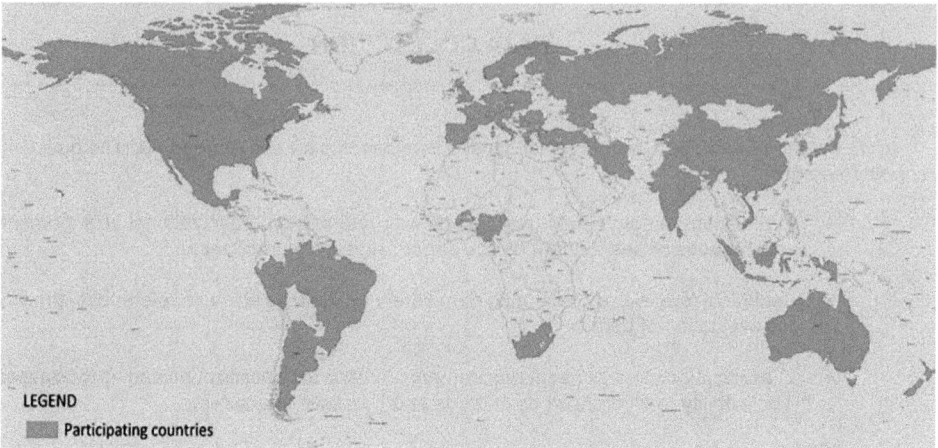

LEGEND

Participating countries

Fig 4.1
Carte du monde – Pays ayant participé à l'enquête

Une vue globale des pays ayant répondu à l'enquête et les principaux enseignements font l'objet de l'annexe B.

Au cours de l'étude, on a noté que les systèmes nationaux ou provinciaux pour la sécurité des barrages font généralement l'objet de révisions et de développements continus. Selon le moment où les données d'un pays spécifique ont été renseignées et vérifiées, sur la période allant de 2009 à 2013, il faut avoir à l'esprit que les éventuelles mises à jour ou changements récents ne sont pas intégrés dans le rapport.

Dans de nombreux cas, les questionnaires ont été renvoyés avec des informations plus détaillées Cela a permis une compréhension plus approfondie de la façon dont ces pays gèrent la sécurité des barrages. Cependant, afin d'obtenir des informations homogènes pour mener une analyse comparative, le groupe de travail a décidé de retranscrire tous les questionnaires renseignés dans un format standard. Les informations de base sont consignées dans des fiches de 2 à 4 pages pour chaque pays (annexe C).

Ces fiches contiennent des informations sur les grands principes de gestion de la sécurité des barrages, y compris l'attribution des responsabilités, les modalités de la surveillance et le cadre juridique. Des chapitres distincts présentent le référentiel technique, le mode de classification des barrages, et un résumé des autres informations transmises. Enfin, les fiches contiennent des références à des publications et des sites Internet pertinents.

Pour assurer la qualité des fiches, les projets de celles-ci ont été transmis aux représentants des pays concernés pour vérification et clarifications. Dans la plupart des cas, mais pas tous, la vérification demandée a été effectuée. Les fiches de synthèse par pays participants sont présentées par ordre alphabétique dans l'annexe C.

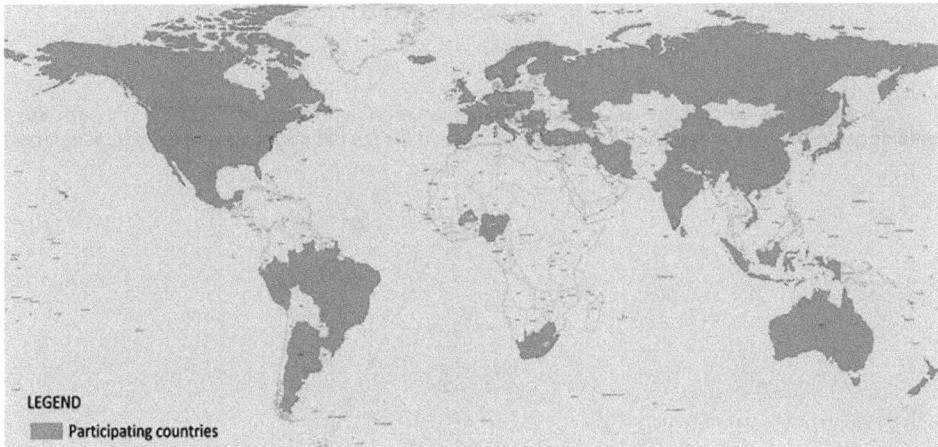

Fig 4.1
World map – participating countries

An overview of the participating countries and the main finding is given in tabular form in Appendix B.

During the study it has been observed that the national or provincial systems for dam safety are commonly subject to review and continuous development. Depending on when the input on a specific country has been documented and verified, over the period 2009 to 2013, it should be noted that the description presented here may not reflect the most recent updates and changes for each country.

The completed questionnaires were in many cases supplemented with more detailed information. This enabled a deeper insight into some of the countries and how they administer dam safety. In order to make the information comparable for further analysis, though, the working group decided to adapt all the completed questionnaires into a standard format. The main information is presented in summaries of 2 to 4 pages for each country.

The summaries contain information about the main principles of dam safety management including assignment of responsibilities, safety supervision arrangements and legal framework. There are also separate sections on technical framework and dam classification, and a summary of any other information given. Finally, the summary reports contain references to relevant publications and websites.

For reasons of quality assurance, the draft summaries were circulated to the respondents in each participating country for clarification and verification. In most cases, but not all, the requested verification has been supplied. The final summary reports of the participating countries are presented in alphabetic order in Appendix C.

5.   RÉSULTATS

L'objectif de ce chapitre est de donner une vue d'ensemble sur la gestion de la sécurité, le contexte réglementaire et la classification des barrages dans les pays ayant répondu au questionnaire.

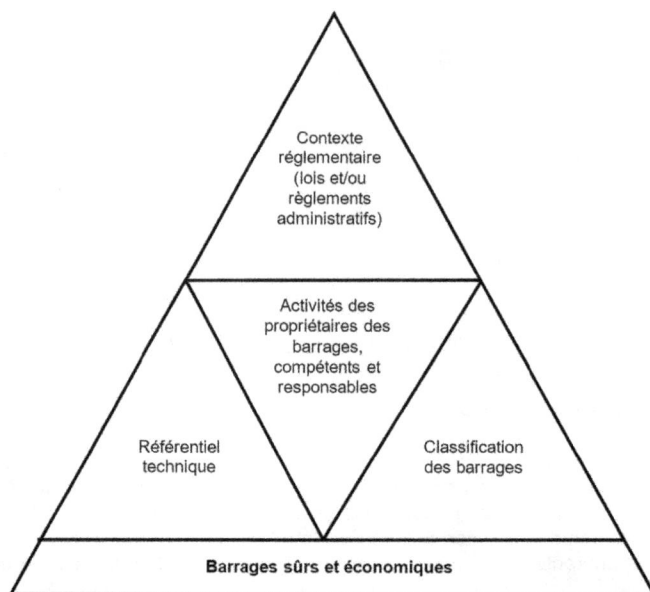

Fig 5.1
Composants de la sécurité des barrages

Le Comité de la Sécurité des Barrages de la CIGB a publié récemment le bulletin 154 "Gestion de la sécurité des barrages en exploitation » [2]. Ce bulletin donne une vue globale sur différentes approches de la gestion de la sécurité et propose des recommandations pour l'élaboration d'un système de gestion de la sécurité. Dans de nombreux pays ayant de grands barrages l'administration d'État a défini des cadres de gestion de la sécurité qui suivent plus ou moins ces recommandations.

Dans ce qui suit le bulletin 154 sert de référence pour la comparaison entre les dispositions recommandées et les systèmes législatifs et les pratiques décrits pour les pays ayant participé à l'enquête en annexe B. Les détails pour chaque pays figurent dans l'annexe C.

Pour plusieurs pays européens, une information résumée sur la réglementation et des liens sur les documents réglementaires sont également disponibles sur le site internet du club européen de la CIGB (http://cnpgb.inag.pt/IcoldClub/index.htm). On y trouve le rapport « Dam Legislation » [3] qui est mis à jour régulièrement au fil des apports des pays participants.

5.1.   PRINCIPES DIRECTEURS DE LA GESTION DE LA SÉCURITÉ DES BARRAGES

Les barrages créent des réserves d'eau pour l'alimentation en eau, la production hydroélectrique, la régulation des cours d'eau, l'irrigation et/ou le contrôle des crues. Bien que ces ouvrages servent l'intérêt général, il ne faut pas oublier qu'ils représentent un danger pour la société s'ils ne sont pas gérés correctement. C'est pourquoi un système de gestion de la sécurité (SGS)

5. RESULTS

The intention of this section is to provide an overview of current practice of dam safety management, dam safety legislation and dam classification in the responding countries.

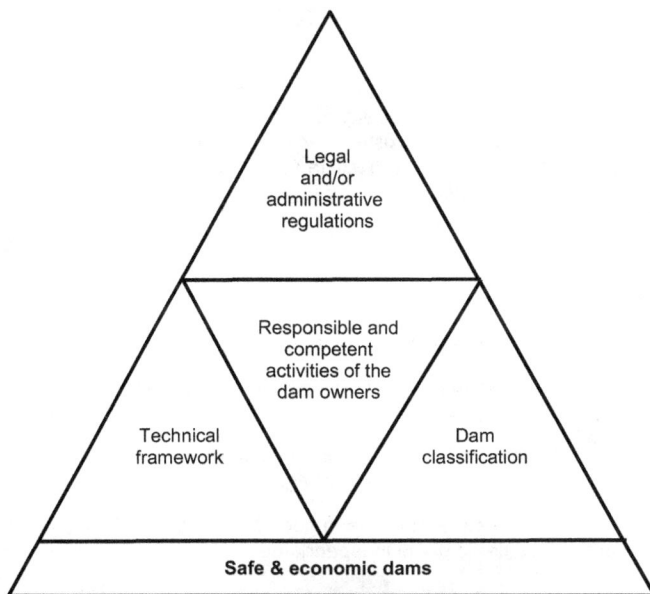

Fig 5.1
Components of dam safety

The Dam Safety Committee of ICOLD has recently prepared Bulletin 154 "Dam Safety Management: Operational Phase of the Dam Life Cycle" [2]. The bulletin gives an overview of different approaches to dam safety management and recommendations for development of a safety management system. In many countries with large dams the state administration has created frameworks for dam safety management which agree more or less with these recommendations.

In the following the ICOLD Bulletin 154 is used as a reference for comparison of recommended arrangements with the current legal systems and practices as reported for the participating countries in Appendix B. Details on each country can be found in Appendix C.

For several European countries summarized information on dam legislation, and links to the actual documents, are also available on the website of the European Club of ICOLD (http://cnpgb.inag.pt/IcoldClub/index.htm). There the report "Dam Legislation" [3] is being revised continuously, when updated information is provided by a participating country.

5.1. MAIN PRINCIPLES OF DAM SAFETY MANAGEMENT

The purpose of dams is to control and store water, for water supply, hydropower production, river regulation, irrigation and/or to provide flood protection. Even though there are public benefits, one must remember that dams and reservoirs also may pose a threat to the society if not properly managed. Thus, it's necessary to have a dam safety management system to protect people, property

est nécessaire pour s'assurer de la protection des personnes, des biens et de l'environnement vis-à-vis des effets néfastes d'une exploitation inappropriée ou d'une rupture des barrages ou de leurs réservoirs. La maîtrise de la ressource en eau est un enjeu d'importance nationale et c'est en général le gouvernement qui est responsable de la promulgation du contexte réglementaire, lois et autres outils réglementaires, pour contrôler ces activités.

5.2.   CONTEXTE RÉGLEMENTAIRE DE LA SÉCURITÉ DES BARRAGES

Le rôle du gouvernement inclut l'élaboration et la promulgation de lois et règlements spécifiques pour la protection du public, des biens et de l'environnement. A peu d'exception près, les pays ayant répondu au questionnaire indiquent qu'il existe, au niveau national ou provincial, des lois sur l'eau faisant référence à la sécurité des barrages. Il existe également des lois sur la gestion de la sécurité et la protection des populations spécifiques aux barrages ou relatives à des installations industrielles similaires à risques importants. Ainsi, dans la plupart des cas, il existe des réglementations minimales traitant de la sécurité des barrages. Des lois et règlements traitant de la sécurité des barrages doivent absolument exister, a minima dans les pays ou régions où sont implantés des barrages dont la rupture auraient des conséquences graves.

Le cadre réglementaire comprend typiquement :

- Une affectation clairement définie des responsabilités pour garantir la sécurité des barrages et l'atténuation des conséquences si un barrage venait à rompre ;

- L'imposition des exigences fondamentales en ce qui concerne la conception, la construction et l'exploitation des barrages ;

- Le principe d'une supervision de la sécurité par le propriétaire ou l'exploitant, et par une autorité ou tierce partie indépendante.

Dans plusieurs pays la classification des barrages est utilisée pour définir clairement ceux pour lesquels la réglementation s'applique, et l'autorité responsable de la supervision (voir section 3.5).

Il existe en pratique deux manières de réglementer la sécurité des barrages :

- Par des lois et règlements s'imposant directement aux propriétaires ou exploitants sans action supplémentaire de l'autorité de contrôle;

- Par des lois et règlements de nature obligatoire aussi bien pour les propriétaires et exploitants de barrage que pour les autorités administratives de contrôle, soit pour s'assurer de la bonne application des lois et règlements par des contrôles officiels, soit pour édicter des règlements contraignants plus détaillés et des autorisations donnant davantage de moyens d'action aux services de contrôle vis-à-vis des propriétaires ou des exploitants.

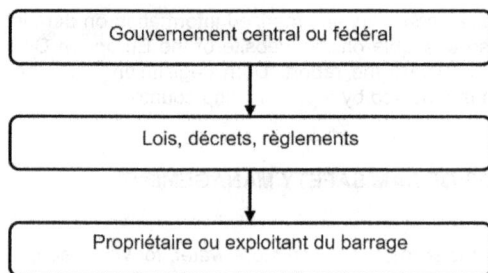

Gouvernement central ou fédéral

↓

Lois, décrets, règlements

↓

Propriétaire ou exploitant du barrage

Fig 5.2
Cadre réglementaire avec imposition directe des textes législatifs "de base" aux propriétaires ou exploitants

and the environment from harmful effects of any improper operation or failure of dams and reservoirs. The control of water resources is a matter of national importance and generally it is the government who is responsible for creating the legal frameworks, laws and other legal instruments to control these activities.

5.2.   LEGAL FRAMEWORK FOR DAM SAFETY

The role of the government includes writing laws and regulations specific to protection of people, property and the environment. With few exceptions the participating countries report that there are national or provincial water acts including references to dam safety. There are also acts on safety management and public protection including references to dams or to comparable structures with significant hazard potential. Also, in most cases lesser regulations on dam safety are in place. Clearly, laws and regulations addressing dam safety should exist, at least in countries and provinces where dams with significant consequences of failure exist.

The legal framework typically includes:

- clear assignment of responsibilities for ensuring the safety of dams, and mitigation of consequences should a dam failure occur

- determination of at least the fundamental requirements for design, construction and operation of dams

- the "four-eye-principle" for dam safety supervision; supervision by the owner or operator and independent supervision by an authority or third party

In several countries classification of dams is used to clearly separate which dams the regulations apply to and not, and which institution (authority) is responsible for supervision of a specific dam, see section 3.5.

In practice two ways exist in order to regulate dam safety:

- Laws and regulations with direct implications for dam owners or operators without supplementary action of state administrative authorities.

- Laws and regulations with obligatory nature for dam owners or operators as well as authorizing authorities or commissioners, either to ensure the execution of the laws and regulations by means of official acting, or to issue more detailed binding regulations and permits with additional force on dam owners or operators.

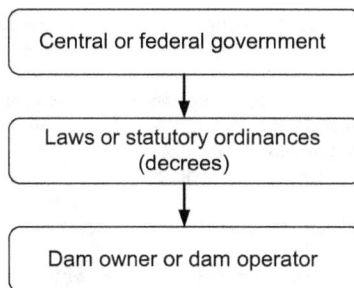

Fig 5.2
Organizational scheme of dam safety regulation with direct implications for dam owners or operators from primary legislation

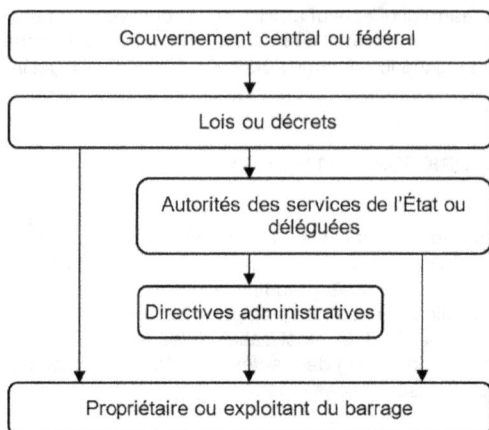

Fig 5.3
Cadre réglementaire avec législation définie au niveau central et application contrôlée au niveau de l'état ou au niveau régional

Les deux méthodes se valent et conduisent à la conception, la construction et l'exploitation de barrages sûrs, ainsi qu'à l'élaboration de plans d'urgence et de mesures de protection des populations.

Pour ces deux méthodes, le contexte des exigences légales peut présenter des caractéristiques différentes :

- Des lois et règlements généraux, tels que des lois sur l'eau, des codes de l'environnement, des exigences pour les activités industrielles et la protection civile, etc. ;

- Des lois et règlements traitant spécifiquement des barrages et de leur sécurité.

La première alternative avec des textes généraux est la pratique la plus commune. La majorité des réponses indique qu'il y a des lois régissant l'usage de l'eau, au niveau national ou régional, traitant des barrages mais pas explicitement leur sécurité. De manière similaire, des lois générales sur la gestion du risque et la protection des populations, applicables aux installations industrielles ou aux activités potentiellement dangereuses pour les personnes ou l'environnement, sont de fait applicables aux barrages et réservoirs en raison de leurs portées. Cependant, ces textes ne traitent pas souvent de manière explicite la sécurité des barrages. On peut d'ailleurs constater que les pays n'ayant pas de législation propre à la sécurité des barrages sont ceux dans lesquels le contrôle administratif est le moins développé (voir figure ci-dessus).

Lorsqu'il n'y a que des textes généraux, sans références aux barrages et à leurs risques, les exigences génériques supposent que le propriétaire (l'entité responsable de la sécurité du barrage) doit exploiter et maintenir le barrage en toute sécurité afin de protéger les populations et/ou l'environnement. De plus, le propriétaire doit disposer des compétences nécessaires pour élaborer et mettre en œuvre les consignes de surveillance et de supervision pour tous les aspects de la gestion de la sécurité du barrage. Cela laisse en pratique une large marge d'appréciation aussi bien aux propriétaires de barrages qu'aux autorités de contrôle. Pour être convenable cette approche présuppose que les propriétaires de barrages soient responsables et compétents (ce mode de fonctionnement par objectif est utilisée en Suède, par exemple).

La deuxième alternative, avec des lois et règlements spécifiques pour la sécurité des barrages, n'est pas aussi répandue. Un tiers des pays ayant répondu à l'enquête possède des textes spécifiques ou des règlements complets pour les barrages et leur sécurité (structures hydrauliques, structures retenant l'eau, réservoirs) mis en œuvre pour protéger explicitement vis-à-vis d'une

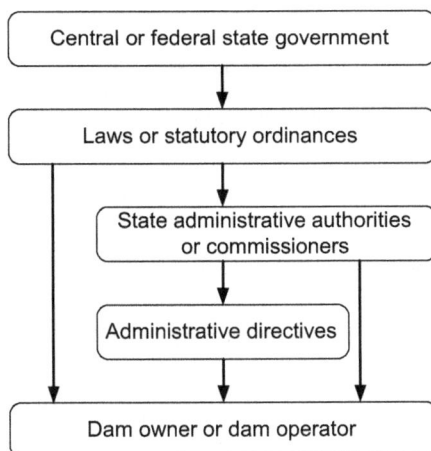

Fig 5.3
Organizational structure for dam safety regulation with centrally controlled legislation and local/state enforcement

Both ways are of comparable value and should address and lead to safe design, construction and operation of dams as well as emergency preparedness and public protection measures.

In both cases the binding legal framework can be of different character:

- Overarching laws and regulations such as water laws, environmental codes, acts for industrial activities and civil protection etc.

- Laws and regulations referring specifically to dams and dam safety.

The first alternative with overarching acts is the more common practice. The majority of the participating countries report, that there is a national or provincial water act in place that includes aspects of dams but not dam safety explicitly. Similarly, general laws on safety management and public protection, with applicability to industrial plants or activities that may harm the public or environment, are likewise valid for dams and reservoirs because of their characteristics. However, they don't often refer to dam safety explicitly. At the same time as acts respectively laws with direct respect to dam safety are reported to be absent in many countries, the same countries commonly have lesser administrative regulations on dam safety in place (see figure above).

Where there are only overarching laws and general regulations, without specific conditions relevant to dams and risks caused by dams, the general requirements imply that the owner (the responsible entity for dam safety) is obliged to operate and maintain the dam safely to protect the public and/or the environment. In addition, the owner should have the necessary competence to draw up and follow appropriate routines for surveillance and supervision for all aspects of dam safety management. In practice this leaves a wide breadth of interpretation for dam owners as well as for the responsible authorities, and to be appropriate this approach presupposes knowledgeable and responsible dam owners. (This type of goal setting framework is used in for example Sweden).

The second alternative, with laws and regulations referring specifically to dam safety, is not as widespread. In about 1/3 of the participating countries there are specific acts or comprehensive regulations on dams and dam safety (hydraulic structures/water retaining structures/reservoirs), that have been established to explicitly protect against misoperation or failure of dams and reservoirs.

mauvaise exploitation ou d'une rupture des barrages et réservoirs. La République Tchèque, la Finlande, la France, la Grande Bretagne, la Norvège, la Slovaquie, la Corée du Sud, la Suisse, les Etats-Unis et plusieurs provinces d'Australie et du Canada en sont des exemples. La portée et le niveau de détail de ces exigences légales et administratives n'ont pas été étudiés ou comparés dans ce rapport, mais sont éminemment variables d'un pays à l'autre.

Plusieurs pays font état de développement en cours de leur système légal pour la sécurité des barrages. Par exemple des lois sur la sécurité des barrages sont en cours de propositions ou ont été élaborées en Suisse (révision), en Grande Bretagne y compris l'Ecosse, le Pays de Galles et l'Irlande du Nord (révision), en Inde, en Suède, en Ukraine et en Roumanie. D'autres pays comme la Finlande et la Norvège viennent de promulguer des textes réglementaires mis à jour.

5.3.   RESPONSABILITÉ DE LA SÉCURITÉ DES BARRAGES

Le bulletin 154 de la CIGB [2] comprend la description du cadre légal et des responsabilités de la sécurité des barrages. Le bulletin expose que la responsabilité première incombe au propriétaire, ou à l'entité responsable dans les cas où le propriétaire du barrage n'est pas une personne, une organisation ou une entreprise unique. Cela implique que le propriétaire du barrage est le responsable ultime de la sécurité des populations, des biens et de l'environnement, autour et à l'aval du barrage. Parfois c'est une institution ou une agence gouvernementale qui est responsable de la sécurité du barrage et des populations.

Les dispositions prises par le propriétaire pour la sécurité doivent être conformes aux exigences et attentes du gouvernement, aux lois et au système réglementaire en vigueur dans le pays ou la région/province où est situé le barrage, quels que soient leur mode d'établissement et de mise en œuvre. L'enquête faite pour ce rapport confirme que, dans la très grande majorité des cas, la responsabilité de la sécurité du barrage incombe bien à son propriétaire.

Environ 20% des pays participants à l'enquête ont signalé que tous ou presque tous les barrages sont propriété de l'État. Dans quelques cas l'entité responsable est un service du gouvernement ou une institution publique doté de compétences d'ingénierie et de sécurité significatives, et qui est responsable de tous les aspects de l'exploitation opérationnelle et de la gestion de la sécurité.

Inversement, l'entité responsable peut ne pas avoir de compétences en ingénierie, mais est néanmoins responsable du respect des obligations fixées par la loi.

Dans peu de cas on mentionne que c'est l'État qui est directement responsable de la sécurité des barrages. Dans ces cas on suppose que l'État est également le propriétaire et qu'il est donc responsable de la gestion de la sécurité du barrage en tant que qu'exploitant, et non en tant que service du gouvernement.

Les responsabilités d'un propriétaire de barrage comportent en particulier :

- L'exploitation, la maintenance et la surveillance appropriées du barrage ;

- L'élaboration d'un plan d'urgence pour minimiser les risques résiduels ;

- Le libre accès à l'ouvrage et aux documents à l'autorité indépendante chargée de la supervision, en général dépendante de l'État, et la fourniture des comptes rendus d'activités et de surveillance ;

- La cartographie des conséquences d'une rupture éventuelle ;

- La responsabilité financière des dommages occasionnés par le barrage en cas d'exploitation inappropriée (avec certaines exceptions en ce qui concerne les actes de guerre, le terrorisme et parfois les catastrophes naturelles).

Examples are Czech Republic, Finland, France, Great Britain, Norway, Slovakia, South Korea, Switzerland, U.S.A., and several federal states/provinces of Australia and Canada. The scope and level of detail in the legal and administrative regulations have not been further studied or compared in this report but vary greatly.

Ongoing development in the field of legal dam safety regulation is reported from several countries in this investigation. For instance, dam safety bills are presently proposed or being implemented in Switzerland (revision), Great Britain including Scotland, Wales and Northern Ireland, (revision), India, Sweden, Ukraine and Romania. Other countries, such as Finland and Norway, have recently implemented revised dam safety acts/regulations.

5.3.   RESPONSIBILITY FOR DAM SAFETY

ICOLD Bulletin 154 [2] includes a description of the legal framework and responsibilities for dam safety. With respect to responsibility for the safety of a dam, the bulletin states that the prime responsibility should rest with the dam owner or the responsible entity in cases where the dam owner is not a single individual, company or organization. This implies that the dam owner is ultimately responsible for assuring the safety of the public, property and environment around and downstream dams. Sometimes a government institution or agency is responsible for the safety of the dam and the public.

The safety arrangements established by the dam owners must conform to the requirements and expectations of government, the prevailing laws and regulatory system of the country/province where the dam is located, regardless of how they are established and implemented. The survey presented here confirms that in the vast majority of cases, the responsibility for dam safety has been clearly assigned to the dam owner.

About 20% of the participating countries have pointed out that all, or almost all, dams are state owned. In some instances, the responsible entity may be a branch of government or a public body with a significant dam engineering and safety management capability, and which is responsible for all aspects of the operational integrity and safety management.

Conversely, the responsible entity may have no engineering capability, but nevertheless carries all responsibility for meeting the intent of the law.

In a few cases it has been reported that the state is responsible for dam safety. In these cases, it is assumed that the state also owns those dams and is thus responsible for management of dam safety in the role as operator and not as a branch of government.

The dam safety responsibilities of the dam owner in particular include:

- Appropriate operation, maintenance and surveillance of the dam

- Emergency preparedness to mitigate residual risk

- Report to, and enable supervision by, an independent party, generally a state authority

- Mapping of the potential failure consequences

- Liability for damages caused by dam failure or improper operation (with certain exceptions with regard to acts of war, terrorism and sometimes natural disasters)

5.4. SUPERVISION DES BARRAGES

La pratique courante au niveau mondial est le principe d'une supervision par « plusieurs regards » (« four eyes principle » en anglais). Cela vaut autant pour les barrages appartenant à des privés qu'à ceux qui sont propriété de l'État ou de collectivités publiques. Le principe fondamental est que le premier regard est celui du propriétaire ou de l'exploitant – qui est responsable de l'auto supervision ou supervision interne – et que le second regard doit être celui d'une autorité indépendante, en général une autorité de contrôle, un panel d'experts, des chargés de mission, ou un panachage de ces différentes entités.

5.4.1. La supervision par le propriétaire

La supervision d'un barrage par son propriétaire ou son exploitant comprend toutes les activités nécessaires pour assurer la sécurité du barrage et les gestes d'exploitation appropriés. Ces activités doivent être consignées dans des instructions opérationnelles détaillées.

Après les premières années d'exploitation, la sécurité d'un barrage est assurée principalement par des activités d'exploitation incluant la surveillance. De nombreux pays signalent qu'il existe des textes réglementaires exigeant que le propriétaire réalise la supervision de son barrage, ce qui comprend l'obligation d'avoir un programme de surveillance formalisé dans des documents. Les résultats de la surveillance doivent également être correctement documentés.

La surveillance des barrages comprend les inspections visuelles, l'auscultation et les contrôles fonctionnels. L'objectif est de détecter et d'analyser les phénomènes visibles et mesurables confirmant le bon comportement du barrage ou indiquant des écarts par rapport au comportement attendu. La surveillance poursuit deux objectifs : la détection précoce des anomalies et la connaissance du comportement à long terme. Dans plusieurs pays le programme de surveillance doit être envoyé au service du contrôle pour approbation, de même que les rapports annuels de surveillance (par exemple en Allemagne, en Grande Bretagne et en Slovénie). Il est fréquent qu'un ingénieur expérimenté et connaissant bien le barrage soit en charge de l'appréciation et du diagnostic des résultats de la surveillance. Cet ingénieur doit normalement avoir un agrément de qualification ou un certificat délivré par le gouvernement.

En complément de cette surveillance systématique, des revues de sécurité périodiques approfondies sont mises en œuvre. Comme souligné dans le bulletin de la CIGB 154 [2], l'objectif principal de ces revues est d'avoir une vision globale du niveau de sécurité actuel du barrage, et de déterminer si des modifications, organisationnelles, managériales ou structurelles, sont nécessaires pour garantir que le niveau de sécurité est correct, et que le principe d'amélioration continue est bien suivi.

La revue de sécurité représente une appréciation complète du barrage en tant que système et fournit une réponse aux questions suivantes :

- Le barrage est-il conforme aux exigences réglementaires en vigueur ?

- Est-ce que les dispositions managériales et organisationnelles sont suffisantes pour maintenir le niveau de sécurité conformément aux exigences ci-dessus jusqu'à la prochaine revue de sécurité ?

Les revues de sécurité sont normalement réalisées périodiquement à une fréquence qui dépend du niveau de risque aux personnes, aux biens et à l'environnement (fonction de la classification qui peut être basée sur les conséquences ou les caractéristiques géométriques des barrages, comme expliqué ci-après au 5.5). Le délai entre chaque revue de sécurité est généralement de 5 à 10 ans, parfois 15, pour les barrages ayant des conséquences potentielles importantes. La revue de sécurité doit être menée par des experts compétents et expérimentés dans le domaine. D'après les réponses à l'enquête, il y a plusieurs manières d'organiser une revue de sécurité, et de désigner le responsable de cette revue, propriétaire ou régulateur, qui fournit les moyens nécessaires. En pratique, pour les barrages ayant des conséquences potentielles importantes, c'est un groupe ou un panel d'ingénieurs qualifiés (ainsi que des experts d'autres disciplines, par exemple des géologues) qui réalise la revue.

5.4. SUPERVISION OF DAMS

All over the world the usual practice of supervision of dams is based on the "four-eyes-principle". This applies to privately owned dams as well as dams in state or public ownership. The fundamental principle is that the first pair of eyes belongs to the dam owner or dam operator – who is responsible for self-supervision or internal supervision – and the second pair of eyes has to belong to an independent body – usually a supervisory authority, boards of experts, commissioners or a mixture of the above mentioned bodies.

5.4.1. Dam owner's supervision

Supervision of a dam by its owner or operator includes all activities which are necessary for permanent ensuring dam safety and appropriate dam operation. Those activities should be established in a comprehensive operational instruction.

After the first years of operation dam safety is mainly ensured by operational activities including dam surveillance. Most countries report that there is a legal requirement for the owner to perform supervision, which includes an obligation to have a documented surveillance program. The results of supervision have to be well documented.

Dam surveillance includes visual inspections, monitoring and functional testing. The aim is to detect and analyze visible and measurable phenomena confirming the performance of the dam or indicating any deviation from the expected behavior. Surveillance serves to both provide early detection of anomalies and to provide knowledge on long term trends of the dam behavior. In several countries the surveillance program has to be sent to the regulator for approval and also yearly surveillance reports (for example in Germany, the UK and Slovenia). Often an experienced dam safety engineer who is familiar with the dam is involved in the assessment and judgment of the results of the surveillance measurements. Such an engineer normally has to have a state approved authorization/qualification or certificate.

In addition to the regular surveillance periodic in-depth safety reviews are carried out. As pointed out in ICOLD Bulletin 154 [2] the main purpose of the safety review is to obtain an overall view of the actual state of safety of the dam, and determine whether any modifications – organizational, managerial or structural – are necessary to ensure that the level of safety is appropriate, and ensure that the principle of continuous improvement is observed.

The safety review should constitute a comprehensive assessment of the dam system and provide answers to the following questions:

- Does the dam conform to current regulatory requirements?

- Are the managerial and organizational arrangements currently in place sufficient to maintain the levels of safety in conformance with the above requirements until the next safety review?

Safety reviews are normally conducted periodically with the frequency depending on the level of risk to people, property and the environment (expressed in consequence or size classification as explained below). The interval for safety reviews is generally from 5–10 or sometimes up to 15 years for large or high consequences dams. The safety review has to be performed by experts with adequate education, experience and expertise. From the survey it is shown that there are different ways to organize safety reviews, and also as to who acts as the responsible party for carrying out (and paying for) the safety review – the operator or the regulator. In practice for large or high consequence dams the review is often made by a group or 'Panel' of qualified engineers (and experts in other fields – e.g. geologists).

Ces experts sont souvent désignés ou approuvés par les autorités adéquates du gouvernement (par exemple en Norvège, Angleterre, Ecosse, Autriche et en Suisse).

Le rôle du régulateur (ou service du contrôle) est expliqué plus en détail ci-dessous.

5.4.2. *Dispositions prises pour une revue de sécurité indépendante*

Dans une vision moderne de la sécurité, la mise en œuvre d'un cadre réglementaire de sécurité des barrages et d'attribution des responsabilités comprend souvent la création d'une entité indépendante chargée d'assurer la sécurité des barrages.

Ce régulateur doit être indépendant vis-à-vis du propriétaire du barrage et des autres parties prenantes afin d'être libre par rapport à des pressions indues. Si l'entité responsable de l'exploitation et de la sécurité du barrage est une branche du gouvernement, celle-ci doit être clairement séparée et effectivement indépendante des branches du gouvernement ayant le rôle de régulateur. Le régulateur doit être investi de l'autorité réglementaire adéquate, avoir les compétences techniques et managériales et disposer des moyens humains et financiers nécessaires pour satisfaire à ses obligations.

La supervision des barrages par une autorité indépendante ou un expert comprend d'abord la vérification de la conformité des actions du propriétaire ou de l'exploitant par rapport aux exigences réglementaires à suivre. Il comprend ensuite la vérification des résultats (ou leur validité) de la supervision de la sécurité par le propriétaire. Le régulateur doit également mener des investigations, ou les faire entreprendre par le propriétaire, pour clarifier des problèmes ou des points incohérents.

L'enquête a montré qu'il y avait encore certains pays où une entité indépendante de supervision n'existait pas.

5.5. CLASSIFICATION DES BARRAGES

La classification des barrages est mise en œuvre au niveau mondial pour définir les barrages qui doivent faire l'objet d'une gestion de leur sécurité, et pour l'élaboration des lois et règlements. Les pratiques de classification, basées sur les conséquences d'une rupture, sur le potentiel de danger, ou simplement sur des critères géométriques (le plus souvent la hauteur du barrage et le volume du réservoir), sont dictées par des considérations économiques, socio-économiques et les exigences sociétales. Une classification en fonction des conséquences ou du potentiel de danger est une façon de graduer le niveau d'exigences réglementaires de sécurité et de protection civile. Cette gradation contribue à assurer un risque très faible (probabilité d'occurrence) de dommages aux personnes, aux biens et à l'environnement sans avoir des exigences de sécurité irréalistes pour les barrages à risque faible.

Il faut noter que les classifications basées sur les conséquences ou sur les risques ne dépendent pas de la probabilité de rupture ou de l'efficacité des mesures de protection civile. La classification par conséquences est obtenue à partir de données caractérisant les vallées en aval (présence de zones habitées, d'infrastructures, d'enjeux environnementaux, etc.).

En classifiant ainsi les barrages on accorde habituellement le plus d'attention à ceux qui ont les conséquences ou le potentiel de danger les plus importants, ce qui permet d'employer au mieux les ressources. La classification des barrages peut servir à définir le périmètre et l'étendue du champ réglementaire, et à fixer des exigences différentes pour la conception, la construction et l'exploitation selon l'importance des conséquences ou du potentiel de dangers. Les classes peuvent servir à définir les sollicitations requises, par exemple les crues, et les charges hydrostatiques associées, varient d'une crue centennale pour les barrages ayant un faible potentiel de danger, jusqu'à une crue décamillénale, ou la CMP (Crue maximale probable), pour un barrage à fort potentiel de danger.

Experts are often appointed or approved by the relevant ministry or authority of the government. (Some examples of countries where experts are officially approved are Norway, England, Scotland, Austria and Switzerland.).

The role of the regulator is further explained below.

5.4.2. Arrangements for independent dam safety review

In terms of the modern view of safety governance the government's establishment of a regulatory framework for dam safety and assignment of responsibilities often includes the establishment of an independent regulatory body to assure the safety of dams.

The regulatory body should be independent from the dam owner and other parties so that it is free from any undue pressure from interested parties. If the responsible entity for operation and safety of the dam is a branch of government, this branch should be clearly separated from and effectively independent of the branches of government with responsibilities for regulatory functions. The regulator should have adequate legal authority, technical and managerial competence, and human and financial resources to fulfill its responsibilities.

Supervision of dams by an independent authority or expert includes primarily the check whether the dam owner or operator works in accordance to the regulations which are to be followed. Secondly it includes the check of the results (or of the validity of the results) of supervision of the dam by the owner. The supervisory body also has the task to make investigations into or to instigate investigations by the owner in order to clear up problems or inconsistencies.

From the survey it has been found that there are still some countries where an independent supervisory body does not exist.

5.5. DAM CLASSIFICATION

Classification of dams is applied throughout the world as a way of defining dams subject to dam safety management and development of laws and regulations. The practice of classifying dams by consequences of failure, hazard potential, or simple geometrical parameters (most often dam height and reservoir volume), arises from economic, socioeconomic and social needs. Consequence or hazard classification is a way of grading the level of safety requirements and civil protection measures enforced by dam safety regulations. This gradation helps to ensure a very low risk (probability of occurrence) of damage to people, property and environment without creating unrealistic safety requirements for low hazard dams.

It should be noted that consequence or hazard classification of dams is independent of the probability of failure, and of the efficiency of civil protection measures. Consequence or hazard classification of dams is achieved by consideration of data for the downstream valley (presence of residential areas, infrastructure, environmental values etc.).

By classifying dams, most attention is usually given to dams with the highest consequences or hazard potential, ensuring the most effective use of resources. Dam classification can be used to limit the scope and extent of dam safety laws and regulations, and to set different requirements for design, construction and operation of dams according to the failure consequences/hazard potential. As an example, dam classes may govern the required level of loads – e. g. floods with their corresponding water levels may vary from a 100-year flood for low hazard dams to a 10.000-year flood or PMF (probable maximum flood) for a high hazard dam.

Le diagramme suivant donne une vision globale des approches de classification et leurs objectifs.

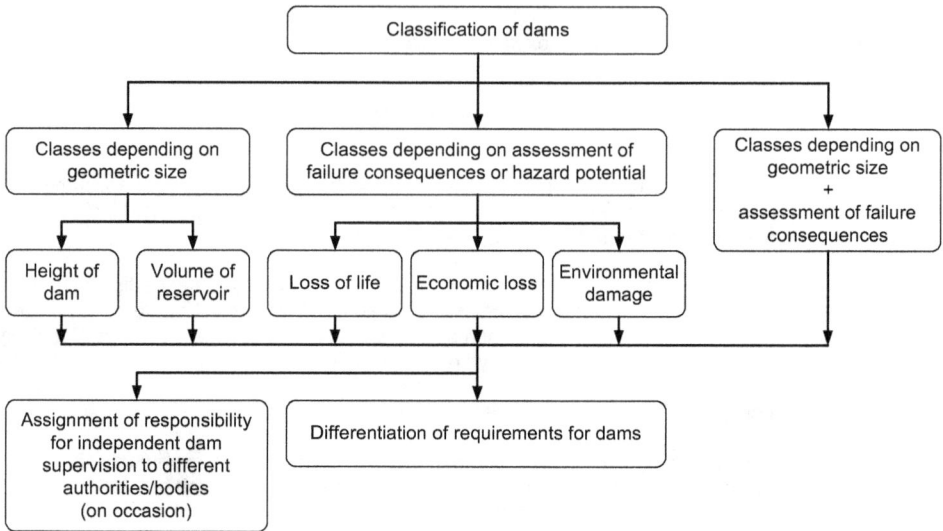

```
                        ┌─────────────────────────┐
                        │   Classification of dams │
                        └─────────────────────────┘
```

| Classes depending on geometric size | Classes depending on assessment of failure consequences or hazard potential | Classes depending on geometric size + assessment of failure consequences |

| Height of dam | Volume of reservoir | Loss of life | Economic loss | Environmental damage |

| Assignment of responsibility for independent dam supervision to different authorities/bodies (on occasion) | Differentiation of requirements for dams |

Fig 5.4

Différentes approches pour la classification des barrages et leurs objectifs

Les résultats de l'enquête sont présentés dans les chapitres ci-après. On trouve des éléments complémentaires pour un certain nombre de pays européens dans le rapport du club européen de la CIGB : « Sécurité des barrages en service » (2012) [4], disponible sur le site web du club européen.

5.5.1. *Exigences légales*

La plupart des pays définissent les critères de classification dans des lois ou des règlements, ou les deux. Dans quelques pays la classification des barrages est utilisée pour spécifier les décrets, règlements ou les exigences légales au sein des décrets et règlements qui sont à appliquer aux barrages de chaque classe. En Norvège par exemple, pour les barrages de la classe de moindres conséquences, seul un petit nombre d'exigences non techniques s'applique.

Dans plusieurs pays la classification des barrages est aussi utilisée pour définir l'autorité en charge de la supervision pour le gouvernement. Dans presque tous les pays fédéraux ayant des états munis d'une souveraineté importante (états avec leurs propres gouvernements, parlements et cour de justice) la classification est utilisée dans ce but. Dans ces pays les autorités fédérales ont en charge certains barrages, tous les autres étant de la responsabilité des états souverains. En principe le gouvernement fédéral est responsable des grands barrages tandis que les états ou régions supervisent des barrages plus petits. C'est le cas par exemple en Autriche et en Suisse. Les critères de classement sont en général basés sur les caractéristiques géométriques de hauteur et de volume du réservoir.

Le questionnaire ne comportait pas de questions spécifiques pour savoir si la classification était utilisée pour graduer les sollicitations, mais quelques pays ont tout de même indiqué ce type d'information (voir annexe C). La classification semble dans ce cas utilisée pour deux critères :

- La crue (ou la tornade) de projet ;

- Le séisme de référence.

The following flow chart provides an overview of approaches for dam classification and its goals.

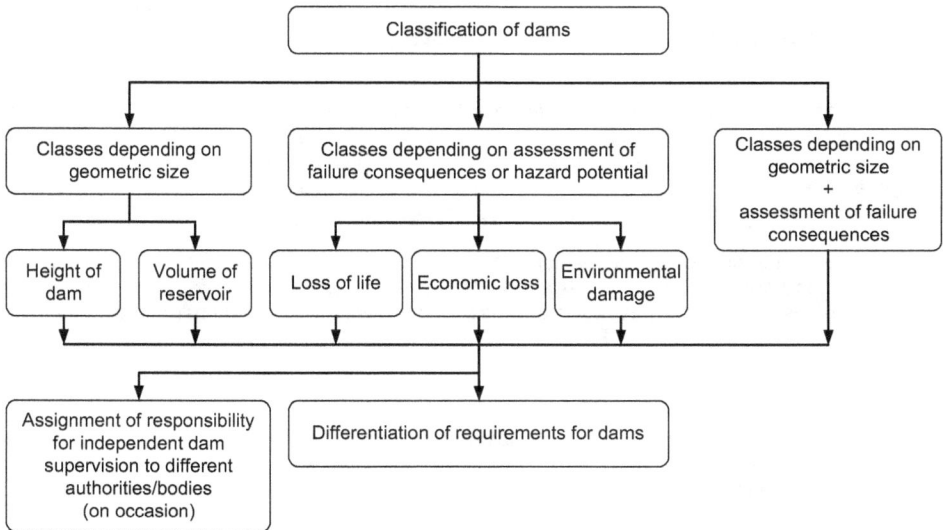

Fig 5.4
Approaches for dam classification and its goals

Findings from the survey are presented in the sections below. (More details on dam classification in a number of European countries can be found for example in the ICOLD European Club report "Safety of Existing Dams" (2012) [4], available on the web site of ICOLD European Club.).

5.5.1. Legal requirements

Most countries have requirements for the classification of dams either in laws or regulations, or in both. In some countries classification of dams is used to specify which acts and regulations, or which legal requirements within acts and regulations, are to be applied to dams of each class. (In Norway, for example, only a few non-technical requirements in the dam safety regulation apply to dams in the lowest consequence class.).

In several countries classification of dams is also the basis for defining which authority is responsible for governmental supervision. In some, but not all, federations with sovereign self-governing states (states with their own government, parliament and courts) classification is used for this purpose. In such federations, administrative authorities of the federal government are responsible for certain dams and administrative authorities of the governments of the single sovereign states are responsible for the remaining dams. Normally, the federal authority is responsible for supervision of the large dams, whereas the single state administrations are responsible for supervision of the smaller dams. (This is the case in, for example, Austria and Switzerland.) The criteria for dam classification are normally the geometrical sizes "height of dam" and "volume of reservoir".

The questionnaire did not specifically include questions about whether dam classes are used to differentiate requirements for design loads etc., but some countries have provided such information, as shown in the Appendix C. Classification seems to be most commonly used for two specific dam design criteria:

- Design flood or design storm respectively.

- Design earthquake.

Les valeurs à retenir pour ces critères de conception dépendent de la classe du barrage, comme mentionné ci-dessus.

5.5.2. Critère de classification et nombre de classes

Les barrages sont répertoriés habituellement en deux, trois ou quatre classes, et plus rarement jusqu'à sept classes. Dans la majorité des pays, on trouve de trois à quatre classes. Les principaux critères retenus pour répartir les barrages en classe sont soit des paramètres géométriques du barrage, parfois combiné avec la taille du réservoir, soit le potentiel de dangers ou les conséquences d'une rupture, soit une combinaison de ces différentes possibilités.

Les paramètres géométriques pour répartir les barrages dans différentes classes sont généralement la hauteur du barrage (exprimée en mètres) et le volume du réservoir (exprimé en m^3) ; les barrages sont ainsi répartis dans des classes qui reflètent la taille du barrage et de celle du réservoir :

- Grands barrages

- Barrages moyens ou intermédiaires

- Petits barrages

On trouve aussi des classes spécifiques pour les "très grands barrages" et / ou les "très petits barrages".

Cette enquête n'a pas étudié les différences entre les conséquences d'une rupture et le potentiel de danger. Les conséquences d'une rupture utilisées pour classer les barrages sont généralement les pertes de vies humaines (exprimés en nombre d'habitations, victimes ou populations potentiellement impactées) et les pertes économiques (dommages matériels, exprimés en coûts). Les impacts sur les enjeux environnementaux, culturels et sociaux sont parfois pris en compte séparément.

Les classes reflètent donc, bien entendu, les pertes potentielles attendues :

- Les barrages ayant des conséquences importantes en cas de rupture, ou avec un fort potentiel de danger ;

- Les barrages ayant des conséquences moyennes ou significatives en cas de rupture, ou un potentiel de danger moyen ;

- Les barrages ayant des conséquences peu importantes en cas de rupture, ou un faible potentiel de danger ;

Quelques pays ont des classes spécifiques pour les barrages ayant des conséquences très importantes ou très sévères, et pour ceux dont les conséquences sont jugées insignifiantes.

La classification à partir de paramètres géométriques a été pendant longtemps la méthode la plus répandue, mais celle à partir des évaluations qualitative ou quantitative des conséquences d'une rupture devient plus fréquente ces dernières années.

Dans plusieurs pays les deux variantes de classification sont combinées. Ainsi, lorsque la classification au moyen de critères géométriques est préconisée, une appréciation qualitative des conséquences de la rupture ou du potentiel de danger est souvent recommandée ou même exigée pour consolider le choix de la classe appropriée pour le barrage.

The selection of values for these design criteria depend on the selected dam class, as mentioned above.

5.5.2. Criteria for classification and number of classes

Dams are usually divided into two to four classes, and infrequently up to seven classes. In most countries three or four dam classes are chosen. The main criteria for dividing dams into classes are either geometrical dam parameters, sometimes combined with reservoir size, or consequences/ hazard potential due to dam failure, or a combination of these.

The geometrical parameters for subdividing dams into dam classes are generally height of dam (expressed in meters) and volume of reservoir (expressed in m³), and the dams may be divided in classes reflecting the dam/reservoir size:

- large dams,

- medium/intermediate dams,

- small dam

Some also have separate classes for "very large dams" and/or "very small dams".

This review has not investigated the difference between consequences of failure and hazard potential. Dam failure consequences used for classification are generally losses of life (expressed in number of dwellings, victims or population at risk) and economic losses (material damages, expressed in money). The loss of environmental, cultural and social values is sometimes considered separately.

Thus, the dam classes reflect, of course, the expected potential losses:

- dams with high consequences of failure/hazard potential,

- dams with medium / significant consequences of failure/ hazard potential and

- dams with low consequences of failure/hazard potential.

Some countries have separate classes for dams with very large/very severe consequences of failure and for dams with insignificant consequences of failure.

The classification by means of quantitative geometrical sizes has been the more usual means whereas classification by means of qualitative or quantitative estimation of failure consequences is becoming more common in recent years.

In several countries both variants of classification are applied in combination. In cases where classification by means of geometrical parameters are preferred, a qualitative assessment of failure consequences or hazard potential respectively will often be recommended or even demanded in order to support the selection of the appropriate dam class.

Dans de nombreux pays les paramètres géométriques, que sont la hauteur du barrage et le volume du réservoir, peuvent à eux seuls refléter (ou même remplacer) le critère basé sur les conséquences d'une rupture. C'est le cas typique des pays et régions dans lesquels les zones en aval ont des caractéristiques uniformes, par exemple une densité de population importante. La rupture de n'importe quel barrage dans ce type de région du monde aura des conséquences catastrophiques. Dans d'autres pays, les conséquences d'une rupture dépendent beaucoup plus de la localisation du barrage que de sa taille ou de celle de sa retenue.

À titre d'exemple, la rupture de la plupart des grands barrages d'Europe aura des conséquences désastreuses en pertes de vies humaines, en dommages aux biens et à l'environnement et doivent donc être classés comme des barrages ayant des conséquences ou un potentiel de danger (très) importants. Dans des zones très reculées des pays scandinaves, avec des occupations humaines dispersées, et parfois inexistantes, la rupture d'un grand barrage peut avoir beaucoup moins de conséquences et être classés comme un barrage ayant des conséquences ou un potentiel de danger moyen. Dans ce dernier cas de figure, une évaluation des conséquences (parfois en combinaison avec des paramètres géométriques) est sans doute une base plus solide pour la classification des barrages qu'une classification uniquement basée sur des critères géométriques.

5.5.3. Systèmes de classification pour différents types de barrage et structures hydrauliques

Pratiquement la moitié des pays ayant répondu signalent que les systèmes de classification par conséquences sont également utilisés pour la classification des autres structures hydrauliques, par exemple les seuils en rivière, les barrages de stériles miniers, les écluses et les canaux, et même dans quelques cas les conduites forcées et les galeries en charge des aménagements hydro-électriques.

5.6. REVUE DES RÉFÉRENTIELS TECHNIQUES

Les lois, ordonnances, décrets ou directives administratives peuvent inclure directement des exigences techniques et opérationnelles. L'alternative qui est la plus communément utilisée est d'inclure l'exigence générique d'utiliser « les meilleures techniques possibles » et/ou « les règles techniques largement reconnues ». Ces règles universelles sont désignées ci-après comme étant le référentiel technique.

Les référentiels techniques pour la sécurité d'un barrage peuvent comporter :

- Des standards nationaux ou des normes ;

- Des guides ;

- Des instructions ;

- Les bulletins de la CIGB.

Agir conformément à ces règles implique que les attendus vis-à-vis de la sécurité du barrage seront respectés. Dans la majorité des pays, la réglementation, fédérale ou des états (lois ordonnances, décrets et directives administratives) exige que les barrages soient conçus, construits et exploités suivant les meilleures règles techniques disponibles ou pertinentes. Dans quelques pays, la réglementation demande la conformité à « l'état de l'art scientifique et technique » pertinent (par exemple la loi sur les ouvrages de rétention d'eau en Suisse). Cette exigence requiert un haut niveau de connaissance des techniques les plus récentes dans ce domaine. Cela signifie que pour chaque site spécifique, et indépendamment du niveau d'exigence du référentiel technique, la sécurité du barrage doit être garantie pour toutes les sollicitations et tous les cas d'exploitation prévisibles.

In many countries and regions, the geometrical parameters "height of dam" and "volume of reservoir" may in itself reflect (or even replace) criteria of failure consequences. This is typically the case in countries and regions where areas downstream of dams are comparatively uniform, for example in densely populated countries. A failure of any dam of a certain size in such a region will lead to major consequences. In other countries and regions, consequences of a dam failure are often highly dependent of the location of the dam, more than the size of dam and reservoir.

As an example, most large dams in Central Europe, will cause extensive consequences to life, property and/or environment in case of failure, and should therefore be considered as (very) high consequence/hazard dams. In very remote areas of the Scandinavian countries, where permanent settlements are scattered, or even non-existing, a failure of a large dam may result in far less consequences and can therefore be considered as a medium consequence/hazard dam. In the latter case, a consequence assessment (perhaps in combination with geometrical parameters) is probably a sounder basis for classification of dams than geometrical parameters alone.

5.5.3. Classification systems for different types of dams and hydraulic structures

Almost 50% of the participating countries have reported that consequence classification systems are also used for classification for other hydraulic structures, for example weirs, tailings dams, sluices and in some cases also penstocks and headrace tunnels for hydropower plants.

5.6. TECHNICAL FRAMEWORK OVERVIEW

State laws, ordinances, decrees or administrative directives may include technical and operational requirements on dam safety directly. The alternative and more common way is to include a universal binding requirement for the dam owner or operator to use "best available technology" and/ or "generally recognized rules of technology". Such universally valid rules are here referred to as the technical framework.

The technical framework for dam safety may consist of:

- National standards or norms,

- Guidelines,

- Instructions,

- ICOLD Bulletins.

Working in accordance to these rules means that the (normal) expectations with respect to dam safety can be expected to be met. In the majority of countries, the federal or state regulations (laws, ordinances, decrees and administrative directives) demand that dams are to be designed, constructed and operated according to the relevant or best available "recognized rules of technology". In some countries state regulations demand accordance to the relevant "state of the art in science and technology" (e.g. Water Retaining Facilities Act of Switzerland). This requirement demands a high level of topicality. For each site-specific case, and independent of the demanded level in the technical framework, dam safety must be guaranteed for all foreseeable operating and loading cases.

Les rédacteurs des règles techniques peuvent penser que leur production représente l'état de l'art dans le champ spécifique de la sécurité des barrages, et c'est effectivement le cas quand les règles pertinentes sont à jour et encore récentes. Ils développent des règles universellement acceptables, mais il peut y avoir avec le temps une perte de leur pertinence. Il est donc très important de garder à jour le référentiel technique et cela conduit leurs rédacteurs à adapter ces règles pour leur mise à jour.

Les référentiels techniques (standards, guides, etc.) sont des recommandations de fait, mais normalement ne sont pas contraignants. Leur application n'est pas obligatoire pour les propriétaires ou les exploitants, mais ceux-ci en tiennent compte pour les cas de problèmes de sécurité (accidents, rupture de barrages).

Le suivi des référentiels techniques peut devenir une obligation s'ils sont cités ou exigés dans les lois ou décrets. C'est la pratique de plusieurs pays/États.

Les rédacteurs de ces référentiels techniques peuvent être :

- Des autorités administratives ;

- Des instituts de normalisation ;

- Les comités nationaux de la CIGB ;

- Les comités techniques de la CIGB ;

- Des associations professionnelles techniques ou scientifiques ;

- Des instituts scientifiques d'Etat ou privés ;

- Des universités ;

- Autres (par exemple des experts reconnus).

Des exemples de toutes ces possibilités figurent dans ce bulletin. Dans les pays ou les états fédéraux qui n'ont pas leur propre cadre technique pour la sécurité des barrages, il est habituel de s'appuyer sur des référentiels techniques pertinents émis par d'autres organisations, pays ou états. C'est une approche raisonnable. Dans ce contexte, une option peut consister à se référer aux bulletins techniques de la CIGB qui reflètent et décrivent la pratique technique internationale.

The authors of dam safety rules may feel that that their products represent the state of the art in science and technology in the special field of dam safety and thus indeed may be true when the relevant rules are up to date and still new. They will develop to universally acceptable rules, but in the course of time there may be a loss of relevance. It is very important to keep the technical framework of dam safety up to date, and it leads to the need of the authors to adapt those rules to keep them up to date.

Technical rules (standards, guidelines etc.) possess in themselves the character of recommendations, but they are normally not legally binding. Their application is not mandatory for dam owners or dam operators, but adherence to the rules will be an important consideration a dam owner in cases of dam safety problems (accidents, dam failures).

The application of technical rules can become a duty of the dam owners or dam operators if they are introduced or demanded respectively by laws or statutory ordinances. This is the practice in several countries/states.

Authors of technical rules for dam safety can be:

- administrative authorities,

- institutes of standardization,

- national committees of ICOLD,

- ICOLD (with its technical committees),

- scientific-technological associations,

- state or private scientific institutes or

- universities

- others (e.g. well known experts).

Examples of all forms are demonstrated in this bulletin. In countries and states of federations that do not have their own technical framework for dam safety, it is common to rely upon the relevant technical rules of other organizations, countries or states. This is considered to be a sensible approach. In this context it could be an option to refer to relevant ICOLD Bulletins which reflect and describe the international technical practice.

6. CONCLUSIONS

Le groupe de travail du Comité de la sécurité des barrages a compilé les pratiques de gestion de la sécurité et de classement des barrages de 44 pays de la CIGB sur la période 2008-2012. Les enseignements majeurs sont résumés ci-dessous :

- Établir et maintenir un cadre pour la sécurité des barrages est une procédure du sommet vers la base représenté dans le diagramme suivant :

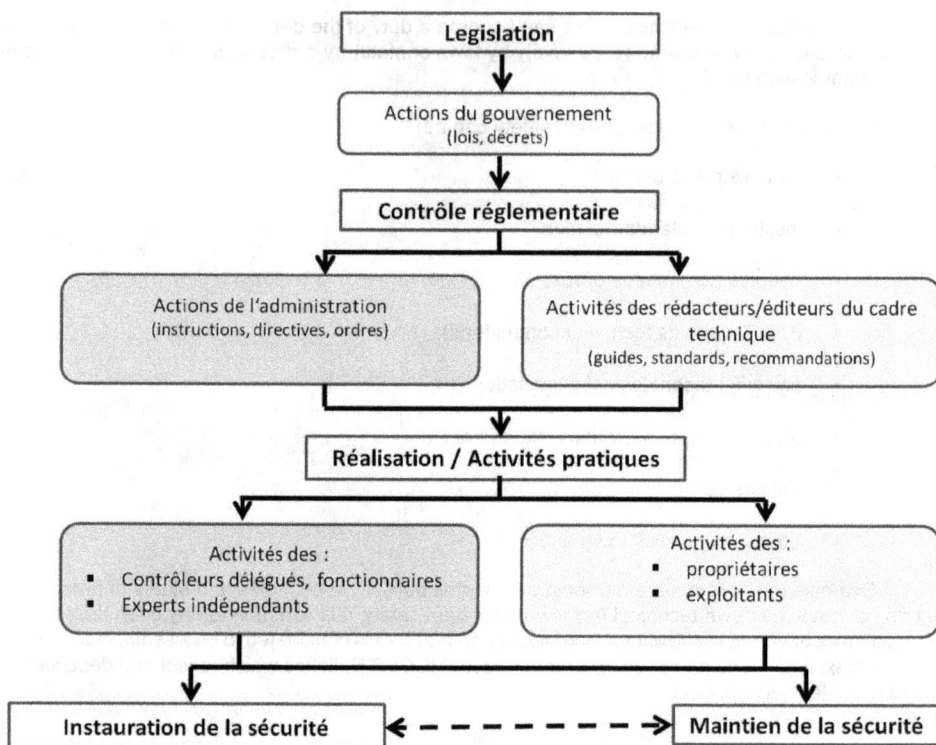

Fig 6.1
Approche pour la mise en œuvre d'un cadre de sécurité des barrages

- Presque tous les pays ont un cadre légal dans lequel les exigences de sécurité sont intégrées soit dans des lois spécifiques de sécurité des barrages, soit dans le cadre réglementaire plus général des lois sur l'eau et de la protection du public. Il existe également des règlements plus détaillés pour la sécurité des barrages ainsi que des référentiels techniques non contraignants.

- Cependant, dans un nombre limité de pays ou d'états/provinces, il n'existe pas de lois sur l'eau, ni autres règlements administratifs et/ou de référentiels techniques exprimant des exigences précises pour la sécurité des barrages. Dans ce cas, on fait souvent référence aux règles et recommandations des autres pays, provinces ou organismes.

6. CONCLUSIONS

The CODS Working Group has collected information about Dam Safety Management and Dam Classification practice in 44 countries from all ICOLD regions in the period 2008–2012. The main findings are summarized below:

- Establishing and maintaining dam safety is a top-down procedure represented in the following flow chart.

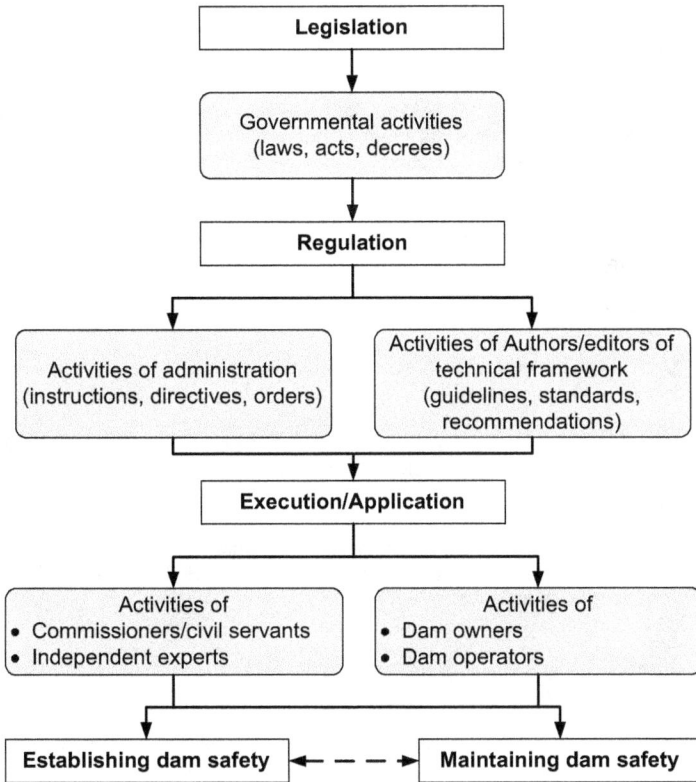

Fig 6.1
Approach for the implementation of a dam safety framework

- Almost all countries have a legal framework where dam safety requirements are included either in laws specifically dealing with dam safety or an overarching legal framework with a water law and a public protection act. In addition, there are normally lower level dam safety regulations and a non-binding technical framework.

- In a limited number of countries and states/provinces, however, water laws, other administrative regulations and/or technical framework with direct requirements about dam safety are absent. In these cases, reference is often given to rules and guidelines of other countries, provinces or bodies.

- Dans pratiquement tous les pays ce sont les propriétaires de barrages qui sont responsables de leur sécurité. Dans quelques pays, la plupart voire la totalité des barrages sont détenus par l'État et dans ce cas-là, la responsabilité de la sécurité des barrages est souvent portée par l'État.

- Dans presque tous les pays la supervision des barrages se base sur le principe d'une double surveillance (auto surveillance par le propriétaire ou l'exploitant, et surveillance indépendante par une entité externe).

- Dans presque tous les pays les barrages sont classés en fonction de leur taille /ou des conséquences de leur rupture ou de leur potentiel de danger.

- Plusieurs pays (environ 50%) ont un classement pour d'autres structures hydrauliques (digues, barrages de stériles miniers, etc.).

- Les méthodes de classification sont très variées. L'emploi de classification basée sur le risque augmente. La majorité des pays qui adoptent ce type de classification utilisent des critères de conséquences, et certains y ajoutent des caractéristiques géométriques.

- Ces approches de classification basées sur le risque qui sont en vigueur dans de nombreux pays, ainsi que le nombre croissant de référentiels techniques utilisant les outils d'évaluation du risque pour la gestion de la sécurité indiquent l'émergence d'une tendance visant à inclure explicitement le risque dans la gestion de la sécurité.

- Dans de nombreux pays la sécurité des barrages engage des experts indépendants (pendant la conception, l'autorisation d'exploiter et la supervision).

- Dans de nombreux pays ayant des grands barrages l'administration et les propriétaires de barrages suivent des procédures de gestion de la sécurité. Le niveau de complexité de ces procédures varie d'un pays à l'autre.

- Les pays où les cadres de gestion de la sécurité des barrages sont peu développés sont encouragés à se référer aux recommandations du bulletin 154 de la CIGB pour sécuriser leurs activités.

- Dam owners are responsible for dam safety in almost all countries. In some countries most of (or all) dams are owned by the state, and in this case the responsibility for dam safety is often with the state.

- In almost all countries the 4-eyes-principle is the basis of dam supervision (self-supervision by the dam owner or operator and independent supervision by a third party).

- In nearly all countries dams are classified depending on their size and/or their failure consequences or hazard potential.

- Several countries (about 50%) also classify other hydraulic structures such as tailing dams, levees etc.

- There is a wide range of approaches to dam classification. There is an increasing application of classification criteria which are risk based. The majority of the countries considered use consequences criteria to classify dams and some with the addition to geometric parameters.

- The approaches to classification by means of risk based criteria in many countries as well as the increasing number of technical guidelines being referenced to dam safety management by means of risk assessment tools indicate an emerging trend to explicitly include risk in dam safety management.

- In many countries independent dam safety experts are involved in achieving dam safety (during the design, permission giving and supervision processes).

- In most countries with large dams the state administration and the dam owners follow certain dam safety management procedures. However, the levels of complexity differ from country to country.

- Countries with less developed dam safety management frameworks are encouraged to refer to the guidelines recommended in ICOLD Bulletin 154 [2] in order to strengthen their activities.

7. RÉFÉRENCES

[1] "Regulatory Frameworks for Dam Safety: A Comparative Study", Bradlow, Palmieri, Salman, World Bank Report, 2002

[2] Bulletin No. 154 "Gestion de la sécurité des barrages en exploitation" ; "Dam safety management: Operational phase of the dam life cycle", CIGB, 2011

[3] "Dam legislation", ICOLD European Club report (continuously updated)

[4] "Safety of Existing Dams", ICOLD European Club report, 2012

D'autres références relatives aux textes réglementaires (lois, actes, ordres, directives, référentiels technique, standards, etc.) sont présentées dans les fichiers de données par pays en annexe C.

7.   REFERENCES

[1] "Regulatory Frameworks for Dam Safety: A Comparative Study", Bradlow, Palmieri, Salman, World Bank Report, 2002

[2] Bulletin No. 154 "Dam safety management: Operational phase of the dam life cycle", ICOLD, 2011

[3] "Dam legislation", ICOLD European Club report (continuously updated)

[4] "Safety of Existing Dams", ICOLD European Club report, 2012

More references with respect to dam safety regulations (laws, acts, orders, directives, technical guidelines, standards etc.) are presented within the Country Data Files included in Appendix C.

ANNEXE A.   QUESTIONNAIRE (TEXTE NON TRADUIT)

APPENDIX A.   QUESTIONNAIRE

The questionnaire included the following questions:

A. Legislation concerning Dam Safety:

1. Do laws, decrees, orders etc. in your country exist or not?

2. **If yes:** Describe briefly the system of a legislation concerning dam safety in your country (laws, decrees, orders, etc.). Present a list of these regulations, their publisher and edition date (date of issue).

3. **If no:** Describe briefly how dam safety is organised.

4. Who is responsible for dam safety? Describe briefly the practice in your country.

5. Are the dam owners or those being responsible for the dams, obliged to have a dam safety surveillance programme?

6. Are there a state / governmental supervision of dam safety in your country? Describe how this supervision is organised.

B. Technical Framework concerning Dam Safety:

7. Do guidelines, standards, norms etc. in your country exist or not?

8. **If yes**: Describe the system of technical framework concerning dam safety in your country (guidelines, standards, norms, etc.). Present a list of these regulations, their publisher and edition date (date of issue).

9. **If no**: Describe briefly how dam safety at a relation to the actual technical state is evaluated.

C. Dam Classification (Categorisation):

10. Does dam classification (categorisation) exist in your country or not?

11. **If yes**: Are the dams classified / categorised according to requirements / criteria given in legislation or technical framework (law, decree, guidelines etc.)? Describe briefly.

12. How many dam classes / categories do you have? How these classes / categories are denoted (marked)?

13. Are there different dam classifications (categorisation) depending on kinds of reservoirs (for instance for "normally water-filled" reservoirs, for "dry" flood protection reservoirs or polders, for weirs, for tailings dams etc.)?

14. Are the dams in your country classified or divided into categories according to risk (hazards and consequences) or other criteria? Describe the principal criteria for division into dam classes / categories; which criteria (loss of life, damage to property, etc) and what are the limits between the different classes/categories (number of lives lost etc.)? If possible, enclose a table showing the different levels / criteria.

15. **If no**: Describe the basis decision used to prescribe the extent of operational dam safety measures (surveillance etc).

ANNEXE B. PAYS PARTICIPANTS – RÉSUMÉ DES RÉSULTATS

No.	Country	Legal or administrative regulation of dam safety		Technical framework		Dam classification according to				Remarks
		yes	no	yes	no	Geometric criteria yes	no	Failure consequences yes	no	
1	Argentina	x		x			x		x	
2	Australia	x* (s)	x*	x			x	x		(s) State acts on dam safety
3	Austria	x		x		x			x	
4	Brazil	x		x		x		x		
5	Bulgaria	x		x		x		x		
6	Burkina Faso	x			x	x			x	
7	Canada	x* (s)	x*	x			x	x		(s) Provincial acts on dam safety
8	China	x		x			x		x	
9	Czech Republic	x		x			x	x		
10	Finland	x (s)		x			x	x		(s) Act on dam safety
11	France	x (s)		x		x			x	(s) Law and decree on dam safety
12	Germany	x*	x*	x		x			x	
13	Great Britain (England, Wales)	x (s)		x			x	x		(s) Reservoirs act
14	Greece	x			x	x			x	
15	Iceland		x		x		x	x		
16	India	x*	x*	x		(x)			x	(x) For design flood only
17	Indonesia	x		x		x		x		
18	Iran	x		x		x			x	
19	Italy	x (s)		x		x			x	
20	Japan	x		x		x			x	
21	Mexico	x		x		x			x	
22	Netherlands	x (s)		x			x		x	(s) Act on flood defences
23	New Zealand	x (s)		x			x	x		(s) Regulations on dam safety
24	Nigeria	x			x	x			x	
25	Norway	x (s)		x		x			x	(s) Regulations on dam safety
26	Peru	x		x		x		x		
27	Poland	x		x		x		x		
28	Portugal	x (s)		x		x		x		(s) Decree on dam safety
29	Romania	x (s)		x		x		x		(s) Law on dam safety
30	Russia	x (s)		x		x		x		(s) Law on dam safety
31	Scotland	x (s)		x			x	x		(s) Reservoirs act
32	Serbia	x		x		x			(x)	(x) Unofficial classification
33	Slovakia	x		x		x		x		
34	Slovenia	x		x		x			(x)	(x) Unofficial classification
35	South Africa	x		(x)		x		x		(x) On selected topics only
36	South Korea	x (s)			x	x			x	(s) Act on dam safety
37	Spain	x (s)		x		x		x		(s) Regulations on dam safety
38	Sri Lanka	x		x		x			x	
39	Sweden	x		x			x	x		
40	Switzerland	x (s)		x		x		x		(s) Law and decree on dam safety
41	Turkey	x		x		x		x		
42	Ukraine	x		x			x	x		
43	USA	x (s)	(x)	x		x	x*	x	x*	(s) Federal laws on dam safety (x) A small no. of states
44	Vietnam	x		x		x		x		

*Différences de réglementation administrative ou légale de la sécurité des barrages dans les différentes provinces/états fédéraux/territoires.

(s) Des lois spécifiques, décrets et/ou des réglementations existent. Des exemples de textes réglementaires sont donnés dans la colonne remarques.

APPENDIX B. PARTICIPATING COUNTRIES
OVERVIEW - SUMMARIZED RESULTS

No.	Pays	Contrôle légal ou administratif de la sécurité des barrages		Référentiel technique		Classification des barrages dépendant de				Remarques
						Critères géométriques		Conséquences d'une rupture		
		oui	non	oui	non	oui	non	oui	non	
1	Argentine	x		x			x		x	
2	Australie	x* (s)		x*			x	x		(s) lois fédérales
3	Autriche	x		x		x			x	
4	Brésil	x		x		x		x		
5	Bulgarie	x		x		x		x		
6	Burkina Faso	x			x	x			x	
7	Canada	x* (s)		x*			x	x		(s) lois par province
8	Chine	x		x			x		x	
9	Rép. Tchèque	x		x			x	x		
10	Finlande	x (s)		x			x	x		(s) loi spécifique
11	France	x (s)		x		x			x	(s) loi et décrets spécifiques
12	Allemagne	x*		x*		x			x	
13	Grande Bretagne (Angleterre, Pays de Galles)	x (s)		x			x	x		(s) lois sur les réservoirs
14	Grèce	x			x	x			x	
15	Islande		x		x	x		x		
16	Inde	x*		x*		(x)			x	(x) pour le choix de la crue de projet uniquement
17	Indonésie	x		x		x		x		
18	Iran	x		x		x			x	
19	Italie	x (s)		x		x			x	
20	Japon	x		x		x			x	
21	Mexique	x		x		x			x	
22	Pays Bas	x (s)		x			x		x	(s) loi de protection contre les crues
23	Nouvelle Zélande	x (s)		x			x	x		(s) lois spécifiques
24	Nigeria	x			x	x			x	
25	Norvège	x (s)		x			x	x		(s) lois spécifiques
26	Pérou	x		x		x			x	
27	Pologne	x		x		x		x		
28	Portugal	x (s)		x		x		x		(s) décret spécifique
29	Roumanie	x (s)		x		x		x		(s) loi spécifique
30	Russie	x (s)		x		x		x		(s) loi spécifique
31	Ecosse	x (s)		x			x	x		(s) loi sur les réservoirs
32	Serbie	x		x		x			(x)	(x) classification non officielle
33	Slovaquie	x		x		x		x		
34	Slovénie	x		x		x			(x)	(x) classification non officielle
35	Afrique du sud	x		(x)		x		x		(x) sur des sujets spécifiques uniquement
36	Corée du sud	x (s)			x	x			x	(s) loi spécifique
37	Espagne	x (s)		x		x		x		(s) règlement spécifique
38	Sri Lanka	x		x		x			x	
39	Suède	x		x			x	x		
40	Suisse	x (s)		x		x		x		(s) loi et décrets spécifiques
41	Turquie	x		x		x		x		
42	Ukraine	x		x			x	x		
43	États Unis	x (s)	(x)	x		x	x*	x	x*	(s) Lois fédérales (x) petit nombre d'états
44	Vietnam	x		x		x			x	

*Differences regarding legal or administrative regulation of dam safety in different provinces/federal states/territories.

(s) Specific acts, decrees and/or regulations on dam safety exist. Examples of legal regulations are given in the remark's column.

ANNEXE C.   FICHIERS PAR PAYS

Cette partie comprend les réponses au questionnaire. Pour chaque pays ayant répondu, les données ont été structurées en quelques pages pour donner une vue globale des principes essentiels du système de gestion de la sécurité. Le groupe de travail a autant que possible vérifié ce document en concertation avec le représentant de chaque pays.

Les textes de ces différents fichiers peuvent être consultés dans cette annexe ; ils n'ont pas été traduits en français.

APPENDIX C.   COUNTRY DATA FILES

This section is based on the answers to the questionnaires. For each responding country the input has been restructured to (in a few pages) provide an overview of the main principles of the dam safety system. The working group has as far as possible verified the document on each country in dialog with the respondent in each country.

1. ARGENTINA

Reporter: Francisco Giuliani

1.1. MAIN PRINCIPLES OF DAM SAFETY MANAGEMENT

The Argentine Republic is constituted as a federation of 23 provinces and the Capital (autonomous city of Buenos Aires).

Dam safety management is held by ORSEP, the Dam Safety Regulator Organization. ORSEP is subdivided in four regions (North, Cuyo-Centro, Comahue and Patagonia) with a central administration at the City of Buenos Aires. A Technical Council, integrated by the four Regional Directors and a President, is the Authority of the National Institution.

There are 110 large dams according ICOLD classification and a significant and larger number of small dams. New large hydro projects are in planning and under construction.

1.1.1. Legal Framework for Dam Safety

Argentina has no specific law concerning dam safety. The National Dam Safety Regulator ORSEP, is responsible for the technical regulatory functions and the supervision of the structural safety of national dams and reservoirs, and is therefore the body responsible for enforcing compliance with the structural norms and standards, as determined by the concessionary contracts drawn up by the Federal Government and private consortia.

ORSEP provides technical assistance to provinces on Dam Safety matters, under requirement and specific agreements. ORSEP was created in 1999 (National Government Decree N° 239/99).

Country	Law/Act concerning dam safety	Decrees etc. concerning dam safety
Argentina	no	yes

1.1.2. Responsibilities for dam safety

National dams are operated under private concessions. The Operators are direct responsible for dams safety and are obliged to take all necessary steps to maintain the physical integrity, the proper functioning, and the safety of the dams and reservoirs.

The private concession contracts require each concessionaire to implement a specific emergency action plan. It must take into account the potential emergencies and failures that could occur and their consequences downstream. These plans include a study of different potential emergencies, actions to be taken, water levels that could be reached and flood wave arrival time at the communities located downstream and warning procedures to be followed.

Provincial governments are responsible for evacuation plans through their civil defense institutions.

ORSEP is the regulatory and control authority on dam safety.

Provincial dams are under their government jurisdiction.

1.1.3. Arrangements for Independent Dam Safety Supervision

The Dam Safety Regulator Organization ORSEP is responsible for the technical regulatory functions and the supervision of the structural safety of all national dams. Operation and maintenance are contracted out to private consortia in the form of concessions. Concessionaires are responsible for monitoring, inspection and to maintain an emergency action plan. Annual reports on dam safety monitoring and technical documents are submitted to ORSEP. Provincial Dams are under provincial government supervision. At the moment the National Government is working on and promoting a Dam Safety Federal Law, which shall cover all large dams existing in the country and the new projects.

1.2. DAM CLASSIFICATION

There is no dam classification.

1.3. TECHNICAL FRAMEWORK OVERVIEW

Dam Safety Guidelines by ORSEP were issued in 2011.

Criteria for basic decisions are based on ICOLD guidelines and follows similar practice as those of Institutions such as Corps of Engineers, Bureau of Reclamation, FEMA-FERC in USA.

1.4. SUPPLEMENTARY INFORMATION

Complementary information on Dams and Dam Safety Management can be found on:

[1] Webpage ORSEP: www.orsep.gob.ar

[2] Webpage National Committee on Dams: www.cadp.org.ar

2. AUSTRALIA

Reporter: Andrew Reynolds / Shane McGrath

2.1. MAIN PRINCIPLES OF DAM SAFETY MANAGEMENT

Australia is a federation with 6 sovereign federal states and 2 self-governing main territories with their own governments and laws/acts and orders/decrees.

2.1.1. Legal Framework for Dam Safety

Governance of dam safety is the responsibility of the individual states and the main territories exclusively. There is no central national Australian act with respect to dam safety. Most federal states have dam safety legislation (Water Acts, Dam Safety Acts). All federal states and territories broadly rely on Guidelines of ANCOLD (Australian National Commission on Large Dams) with some having their own specific technical requirements, acts and legal instruments. Because of the importance and meaning of tailings dams in several federal states of Australia special acts (Mining Acts, Environmental Protection Acts) deal with such plants. The federal states and main territories without their own dam safety regulations rely on Guidelines of ANCOLD or other appropriate guidelines.

The following table gives an overview of dam safety legislation in Australia:

State/Federal State/ Province/Territory	Law/Act concerning dam safety	Decrees etc. concerning dam safety
Australia (Federation)	No	No
Queensland (QLD)	yes/d	Yes
New South Wales (NSW)	yes/d	Yes
Victoria (VIC)	yes/i	Yes
Tasmania (TAS)	yes/d	Yes
South Australia (SA)	no	No
Western Australia (WA)	no	No
Northern Territory (NT)	no	No
Australian Capital Territory (ACT)	yes/d	Yes
d - direct demands regarding dam safety;		
i - indirect demands regarding dam safety (for instance by requiring accordance with state of the art or by referring to special decrees etc.)		

2.1.2. Responsibilities for Dam Safety

The dam owners are legally responsible and liable for dam safety.

2.1.3. Arrangements for Independent Dam Safety Supervision

State regulators are appointed for oversight of the application of the dam safety regulations in the federal states and main territories. Assessment committees, senior officers, independent statutory boards or dam safety engineers (experts and advisers) can be engaged for this task depending on the respective state and dam owner.

For instance, in Tasmania an independent Assessment Committee on Dam Construction exists for advising the responsible Minister on dam safety and other matters for new dams. The Committee consists of 6 members of various disciplines and issues construction approvals. For existing dams, the Manager of the Water Management Branch, who is delegate of the responsible Minister, can direct a dam owner to carry out investigations, provide safety reports and carry out remedial works. A Dam Safety Engineer reviews the reports on work carried out by engineering consultants on behalf of the dam owners.

2.2. DAM CLASSIFICATION

2.2.1. General Principles of Classification

In all federal states and main territories dams are divided into classes. The ANCOLD Guidelines form the basis for classification in principle. There are minor variations between the states and territories. The classification systems are generally based on potential failure consequences. The probability of failures has no influence on the classification. In some states the required safety levels of dam structures are based on the need to reduce the risks to tolerable levels. The population at risk (PAR) or the potential life loss (PLL) is used as criteria for classification additionally.

The same classification system will be applied for dams with permanent water impoundment and for dry flood retaining reservoirs. For tailings dams and ash dams specific regulations are often applied.

2.2.2. Dam Classes Overview

The general scheme for subdivision of dams into 7 categories is:

Category	Potential failure consequences
1	very low
2	Low
3	Significant
4	high C
5	high B
6	high A
7	extreme (for instance > 100 PAR)
Note: There is a difference between the ANCOLD Hazard Classifications and the regulatory approaches of the states and territories apart from Queensland.	

2.3. Technical Framework Overview

There are central national Dam Safety Guidelines published by ANCOLD. Different guidance documents by state or territory governments exist additionally. These guidelines are published by the governmental administration itself or by State or Territory Dam Safety Committees or Panels of Experts.

ANCOLD guidelines:

- Guidelines on Concrete Faced Rockfill Dams (1991)

- Guidelines on Strengthening and Raising Concrete Gravity Dams (1992)

- Guideline Supplement on ICOLD Bulletin No. 75 - Roller Compacted

- Concrete for Gravity Dams - 1989 (1991)

- Guidelines on Selection of Acceptable Flood Capacity for Dams (2000)

- Guidelines for Design of Dams for Earthquake (1998)

- Guidelines on Tailings Dam Design, Construction and Operation (1999)

- Guidelines on the Consequence Categories for Dams (2012)

- Guidelines on Dam Safety Management (2003)

- Guidelines on Risk Assessment (2003)

- ANCOLD's Hazard Rating Spreadsheet in Electronic Format (2008)

- Environmental Guidelines (2000).

Guidelines of Department of Environment and Natural Resources of Queensland:

- Guidelines for Failure Impact Assessment of Dams, NRW April 2002

- Queensland Dam Safety Management Guidelines, February 2002

- Guidelines on the Assessment of Tangible Flood Damages, NRW September 2001
  Guidelines on Acceptable Flood Capacity for Dams, NRW February 2007

- Technical Guidelines for the Environmental Management of Exploration and Mining in Queensland, DME January 1995

- Manual for Assessing Hazard Categories and Hydraulic Performance of Dams, EPA 2007 (located in [2])

Guidelines of the Dams Safety Committee of New South Wales:

- DSC1 – General Information, April 1988

- DSC2 – Role, Policies and Procedures, August 1999

- DSC3 – Glossary, April 1998

- DSC5 – Advice on Legal Matters for Dam Owners, August 1996

- DSC11 – Acceptable Flood Capacity for Dams, June 2002

- DSC12 – Operation, Maintenance and Emergency Management Requirements for Dams, April 2003

- DSC12-1 – Addendum to DSC12 Operation, Maintenance and Emergency Management Requirements for Dams, April 2003

- DSC13 – Consequence Categories for Dams, March 2002

- DSC14 – Requirements for Submission of Information by Dam Owners August 2000

- DSC15 – Requirements for Surveillance Reports, January 2003

- DSC16 – Requirements for Earthquake Assessment of Dams, February 2000

- DSC17 – Requirements for Assessment of Retarding Basins, August 2000

- DSC18 – Dam Design and Construction Issues Requiring Particular Consideration, June 2003

- DSC19 – Tailings Dams, November 2005

- DSC32 – Notes on the Administrative Role of the Dams Safety Committee in the Granting of Mining Leases and Approval of Mining Applications, June 1998

- DSC33 – Mining in Notification Areas of Prescribed Dams, June 1998

- DSC34 – Typical Monitoring Programme Requirements for Mining near Prescribed Dams, August 2000

- DSC35 – Mining Contingency Plans to Minimise Loss of Stored Water from Dams, June 1998.

This suite of guidance sheets is currently undergoing a major revision to introduce the concept of tolerable risk – primarily public safety risks – to the safety management of dams. A document Risk Management Policy Framework for Dam Safety was endorsed by the NSW Government in August 2006 and is available on the DSC web site.

Guidelines of Governmental Administration of Tasmania:

- Guidelines for Construction of Earth-fill Dams 2008

- Guidelines for 5 Year Surveillance Reports 2008

- Guidelines for Dam Safety Emergency Plans 2008

- Guidelines for Pre-Construction Reports 2009

- Guidelines for Work as Executed Reports (not on the website as yet)

- Dam Assessment Report Templates 2008

Guidelines of Governmental Administration of Victoria:

In Victoria the ANCOLD guidelines are to be applied for public dams. Special regulation and guidelines exist for farm dams (irrigation and commercial farm dams).

- Strategic Framework for Dam Safety Regulation 2012

- Guidance Note on Dam Safety Decision Principles 2011

- Your Dam Your Responsibility 2007

- Dam Safety Emergency Plan Flipchart 2007

- Consequence Screening Tool Guidelines 2012

- Consequence Screening Tool for Small Dams 2012

2.4. SUPPLEMENTARY REFERENCES

More information about dam safety management, legislation and classification can be found in the following internet sources.

[1] www.damsafety.nsw.gov.au

[2] http://www.ehp.qld.gov.au/land/mining/guidelines.html

[3] http://www.derm.qld.gov.au/water/regulation/referable_dams.html

[4] www.water.vic.gov.au/governance/dam-safety-management

[5] www.dpipwe.tas.gov.au/inter.nsf/themenodes/lbun-945vq9?open

3.   AUSTRIA

Reporter: Elmar Netzer / Pius Obernhuber

3.1.   MAIN PRINCIPLES OF DAM SAFETY MANAGEMENT

Austria is subdivided into 9 provinces. Water Authorities are the Provincial Governors and the Federal Minister of Agriculture and Forestry, Environment and Water Management. The jurisdiction depends on the height of the dam, respectively the storage capacity of the reservoir.

3.1.1.   Legal Framework for Dam Safety

Construction and operation of dams are regulated by the Austrian Water Law. This law stipulates a few requirements concerning dams. First of all, public interests – especially with regard to dam safety – and rights of third parties must not be violated by the construction and operation of dams. The law also states that the dams have to meet the current state of the art and if they don't, they have to be adjusted accordingly.

The following table gives an overview of dam safety legislation in Austria:

State/Federal State/ Province/ Territory	Law/Act concerning dam safety	Decrees etc. concerning dam safety
Austria (Federation/central State)	yes	yes[1]
Provinces[2]	no	no
[1] Specific regulations are edited by the Austrian Commission on Dams (see below). This commission of experts performs administrative duties.		
[2] Provinces are involved in the dam safety management process by appointed experts of the Provincial Governments.		

Further requirements are specified for dams higher than 15 m above foundation level or impounding reservoirs of more than 500,000 m³. For such dams it is required that a project for the construction of a new dam has to be checked by the Austrian Commission on Dams.

This commission consists of experts of all disciplines involved in the construction and supervision of dams. It is established in the Federal Ministry of Agriculture and Forestry, Environment and Water Management as an advisory body for the Water Authorities. The commission also issues guidelines. Furthermore, a subcommittee of the commission carries out the five-year inspections and reports to the whole board once a year about the findings.

3.1.2.   Responsibilities for Dam Safety

The owners respectively the concession holders are responsible for dam safety. For the operation of dams of the category defined above, the owner has to appoint a Dam Safety Engineer (DSE) and several deputies. The DSE and his deputies have to be qualified engineers. They must belong to the owner's organization and have to be vested with the appropriate executive power for taking all measures necessary for the safety of the dam. Furthermore, the DSE or one of his deputies must be available within due time. Smaller companies may get the permission to appoint a DSE from outside their own organization. The required qualifications for the DSE are, in general: Civil engineer at university level and a minimum of 10 years (deputy 5 years) relevant experience. In addition, the DSE has to be familiar with the respective dams.

3.1.3. *Arrangements for Independent Dam Safety Supervision*

The Water Authorities have to verify that the owner makes the necessary provisions for safety. For that the Provincial Governors appoint a Dam Supervisory Officer and a Federal Dam Supervisory Department is established in the Federal Ministry of Agriculture and Forestry, Environment and Water Management.

The Dam Safety Officer takes part in the annual site inspection, which is carried out by the DSE. Personnel from the Federal Dam Supervisory Department together with independent experts execute the five-year inspections. Both, the Dam Safety Officer and the Federal Dam Supervisory Department also check the annual reports prepared by the DSE.

3.2. DAM CLASSIFICATION

Austria's dams have not been classified with regard to the consequences of a failure. The only classification is with regard to the height of the dam, respectively the storage capacity of the reservoir.

The system for ensuring dam safety is compulsory for all dams with height H > 15 m or storage capacity I > 500,000 m³. It can be extended to smaller dams – e.g. because of difficult foundation or unusual construction procedures – by notice of the Water Authority.

3.3. TECHNICAL FRAMEWORK OVERVIEW

The national technical framework concerning dams and dam safety consists of Austrian National Standards with respect to general aspects of civil engineering and Guidelines of the Austrian Commission on Dams. In addition, the German National Standards (DIN 19700, Dam plants; see chapter Germany) is also employed in Austria to some extent.

Guidelines of the Austrian Commission on Dams are, among others:

- Guidelines for the Assessment of Safety against Floods

- Guidelines for the Seismic Analysis of Dams

- Guidelines for Safety Assessment of Embankment Dams

- Guidelines for Central Control Stations which Operate Reservoirs

- Specifications for the Dam Safety Engineer of "Small Dams"

- Handbook: Operation and Supervision of "Small Dams" with longer lasting impoundment

3.4. SUPPLEMENTARY REFERENCES

More information about dam safety management, legislation and classification can be found in the following references.

[1] Heigerth, G., et al.; Assessing and Improving the Safety of Existing Dams in Austria, Q. 68, R. 58, ICOLD Congress Durban (1994)

[2] König, F. & Schmidt, E.; Construction of Dams in Austria: Authorization Procedure, Dams in Austria, published by the Austrian National Committee on Large Dams (1991)

[3] Melbinger, R.; The Austrian Approach to Dam Safety: A Symbiosis of Rules and Engineering Judgment, Int. Symposium on Dam Safety, Barcelona (1998)

[4] Obernhuber, P.; Dam Safety Management in Austria, Proceedings of the Annual Meeting of the Australian National Committee, Sydney (2006)

4.   BRAZIL

Reporter: Fabio de Gennaro Castro

4.1.   MAIN PRINCIPLES OF DAM SAFETY MANAGEMENT

Brazil officially the Federative Republic of Brazil, is constituted by the union of 26 states, and a Federal District, Brasilia. Dams exist in all states.

4.1.1.   Legal framework for dam safety

The federal law - Lei nº 12.334 of September 20, 2010, establishes the national politics for dam safety. Both water and tailings dams are involved in the law, which created the dam safety national information system.

4.1.2.   Responsibilities for dam safety

The operation of dams in Brazil can be done either by private or governmental entrepreneurs. The entrepreneur is responsible for the dam safety.

4.1.3.   Arrangements for independent dam safety supervision

Dam safety activities are supervised by the authorities that authorized the dam operation.

According to the federal law - Lei nº 12.334 of September 20, 2010, that establishes the national politics for dam safety, the national water agency – ANA (*Agência Nacional de Águas*) is in charge to establish and manage the National System of Dam Safety Information, coordinate the authorities in charge of the dam safety supervision, and gather the reports on the safety of each dam.

4.2.   TECHNICAL FRAMEWORK OVERVIEW

Governmental entities publications:

- A translation of USBR's SEED (Safety Evaluation of Existing Dams) program, sponsored by Eletrobrás (Brazilian Energy Utility) in 1987, was remarkable as a way to organize the dam safety procedures in the country.

- A Manual for Dam Safety and Inspection (Manual de Segurança e Inspeção de Barragens) was issued in 2002 by the Ministry of National Integration.

CBDB (Brazilian Committee on Dams) publications:

- Directives for Inspection and Safety Evaluation of Dams in Operation (Diretrizes para a Inspeção e Avaliação de Segurança de Barragens em Operação), Rio de Janeiro, 1983.

- Dam Safety Basic Guide (Guia Básico de Segurança de Barragens), São Paulo, 1999.

4.3.   DAM CLASSIFICATION

The federal law - Lei nº 12.334 is valid for dams that accomplishes at least, one of the following requirements:

- Height of dam $h \geq 15$ m

- Reservoir Volume $V \geq 3 \times 106$ m³

- Reservoir of hazardous materials, according to appropriate legislation

- Medium or high-risk category.

The first two requirements are obvious and depend only on the geometric characteristics of the dam.

The retention of hazardous materials can pollute the soil and the water table, and specific legislation is applied in these cases.

The risk category can be large, medium or small, as defined by the Authorities that authorized the dam operation. The risk category of a dam is based according to its technical characteristics, maintenance status, and conformity to the dam safety plan. It is also a function of the potential life loss, and the degree of environmental, social and economic impacts.

5. BULGARIA

Reporter: Dimitar Toshev

5.1. MAIN PRINCIPLES OF DAM SAFETY MANAGEMENT

Republic of Bulgaria is a unitary state divided into 263 municipalities. The operation of large dams is controlled by Ministry of environment in conformity with respective ministries (Ministry of Economy, Energy and Transport, Ministry of Regional Development, Ministry of Agriculture). Utilization and preservation of water resources and water systems is however controlled by the Ministry of Environment and Waters and by the Basin Directorates.

More than 2,000 dams and tailings dams have been built up in Bulgaria with 216 of them high more than 15 meters. The number of dams is growing with planning of new dams and water-power facilities.

5.1.1. Legal framework for dam safety

There are two laws that set the basis for safety of water construction works in Bulgaria. The legislation concerning safety of water construction works is the following:

- The territory Management Act (TMA) which deals with reliability of study, design and construction activities, updated in July 2003,

- Water Law which sets the rules for safe operation of water engineering facilities and dams, MRDPW 7/2009.

Each of the above two laws is supported by series of regulations and ordinances arranging application of the respective law in the practice.

- Design criteria, documents, scope and content of studies for three design stages (preliminary design, engineering design, detailed design and executive plans) and other requirements concerning design and investment projects (for dams) in Bulgaria are regulated in Ordinance Nr. 4 –Scope and content of investment projects; MRDPW 1/2001 and 5/2001, which enforces the TMA.

- Terms and order of performance of engineering operation and appurtenant structures is defined by Ordinance Nr.3; Normative and design document 4/2005

- Ordnance for essential requirements to construction projects and appraisal of building products; MRDPW 12/2001

- Ordinance Nr.2 for design of structures In earthquake regions MRDPW 10/2007

- Regulation for correct and safe operation and maintenance of facilities for water engineering infrastructure; Official gazette 97/2004

- Regulation for engineering operation of power plants and networks; Official gazette 81/2000

State/Country	Law/Act concerning dam safety	Decrees etc. concerning dam safety
Bulgaria	yes	yes

5.1.2. Responsibilities for dam safety

The dam owners are responsible for the dam safety (from design phase to operation phase/ until decommissioning). Dam owners can be private (persons/companies), municipalities or State.

5.1.3. Arrangement for independent dam safety supervision

Overall dam safety concept is based on three pillars:

- Design safety

- Quality of construction and structural safety

- Surveillance, monitoring, and maintenance

The design criteria, documents, scope and content of studies for tree design stages (preliminary design, engineering design; detailed design and executive plans) and other requirements concerning design and investment projects (for dams) are defined in different Ordinances and Standards.

Quality of construction is regulated Ordnance for essential requirements to construction projects and appraisal of building products; MRDPW 12/2001

Appropriate operation, regular monitoring and control, maintenance and timely rehabilitation are the three basic pillars of the dam safety system.

The terms and order for performance of engineering operation of dams and appurtenant structures are defined by Ordinance Nr. 3. After a large dam is finished, it becomes subject of surveillance. The surveillance and review of performance of a dam is organised in multiple levels of surveillance and review (dam guardian/concessionaire, ministries, state supervision of group of experts) and allows to minimize the risk of having an undetected critical feature or a potential hazard.

5.2. DAM CLASSIFICATION

5.2.1. General principles of classification

Dams, according to Standards for Dams and Water-Engineering Facilities Designing (1985), are classified in four classes (I, II, III and IV) depending on the building material (concrete and reinforced-concrete or embankment), on the type of geological base (rock, sandy-gravelly or weak clay and fine sand) and the height thereof above the foundation elevation. High dams are those of I class. The IV class covers temporary and auxiliary facilities like diversion cofferdams and tunnels used in the time only of dam erection. In addition, dams pass into higher class depending on the consequences at eventual failures (loss of human lives, infrastructure, environment damages and economic losses).

Dam owners are responsible for classification when planning a new dam or reconstruction of existing dam, and for existing dams.

5.3. TECHNICAL FRAMEWORK OVERVIEW

Technical framework that concerns dam safety - Guidelines published by the State Office of Energy are:

- Standards for design of water engineering facilities that defines basic rules was issued by Ministry of construction and town arrangement and Ministry of Energy, 11/1985.

- Bulgarian Code for Design of Embankment Dams 1/1986.

- The Engineering Manuals of the U.S. Army Corps of Engineers, though not binding, are widely used for design of dams during the last couple of years.

- Standard for design of buildings and structures in earthquake regions; Committee of territoritorial and town arrangement; 1987.

- Standard for design of buildings and structures in earthquake regions; Section II: Normative design documents; 3/2003.

- Standards for design of concrete and reinforced-concrete structures, Committee of territoritorial and town arrangement; 1988.

- Standards for design of concrete and reinforced-concrete structures of water-engineering facilities, Committee of territoritorial and town arrangement; 1988.

- Standards for loading and impacts of waves, wind, ice and vessels on water-engineering Committee of territorial and town arrangement; 1988.Ordinance No. 3 about fundamental design principles for design of structures on building projects and impacts thereon (2005).

- Ordinance No. PD-02-20-2/2012 about the design of buildings and facilities in earthquake regions.

In the context of European integration, the Bulgarian standards and regulation shall be replaced gradually by Eurocodes.

6. BURKINA FASO

Reporter: Eloi Somda

6.1. MAIN PRINCIPLES OF DAM SAFETY MANAGEMENT

Burkina Faso is divided in 13 regions and in 45 provinces. The regions are headed by Governors and there are around 8,000 villages. The total population is around 16 million inhabitants.

The annual average precipitations are in order to 800 mm and limited to three or four month per year. That is why there is much dryness in the country. The principal solution to mobilize water for people in towns and in the villages is to construct dams.

There are more than 1,000 registered dams in Burkina Faso, and around 15 of these are dams higher than 15 meters. The number of dams is increasing due to ongoing registration of existing dams and planning and construction of new projects.

6.1.1. Legal framework for dam safety

Burkina Faso doesn't have a specific regulation or legislation related to dam safety. The Government with the support of the Millennium Challenge Corporation of the USA has undertaken studies for the implementation of a Dam Safety Unit for addressing dam safety activities in Burkina Faso. The study will be completed by 2014.

Dam safety is addressed at the moment through several laws and regulations:

- The water law N° 002 – 2001/AN

With the framework of the Water Law dam construction is submitted to authorization to be delivered by the ministry of Agriculture and Water resources after Consultation of the National Council for Water and the technical committee of the Council.

This law also stated the need to ensure that Hydraulic structure and Dam are designed and built taking into consideration the public safety.

- The Land use and management reform law n° 014/96/ADP

Under this law all Dams and reservoirs are considered as public infrastructures and should beneficiate of safety measures.

- The Environmental Code law n°005/97/ADP

The article 2 of this law required that dams and reservoirs should be designed, built and operated to avoid threat on human establishment and the environment.

The article 16, 17 and 24 require the full environment and social impacts assessment and management for dams and reservoirs.

- For Large Dams, the World Bank regulations BP 437 and OP 437 are applied.

Since 1995, all large dams developed have been submitted to the World Bank guidelines with the designation of an International Board of Experts for the safety aspects of Dams from the planning process to the first impoundment of the reservoir.

For the main large dams, a 10 years safety assessment used to be undertaken for the evaluation of the dam safety.

6.1.2. Responsibilities for dam safety

The water resources management including the planning and implementation of dam in Burkina Faso is under the responsibility of the Ministry of Agriculture and water resources. But several entities are planning and building dams. Among these entities we have other public departments like the Ministry of Public works and some NGOs for small dams and small reservoirs. There are a few small dams owned by the mining industry and some private persons. All these dams have required Authorization of the Ministry in Charge of water resources.

For Large dams an agreement is generally signed between the Ministry of Water and the operators of these dams. These Operators are for Hydropower dams the Public Company producing and supplying electricity SONABEL. For Water supply Dams the agreement is signed with the Public Water supply company ONEA. Some dams are operated by private company under also an agreement with the Ministry of water. The sugar company SOSUCO operates a large dam built for the development of the sugar industry.

Under these agreements these operators are responsible for the operation and maintenance of these dams including safety aspects.

The Ministry of Water is responsible to ensure that these dams are operated and maintained according to modern practice and a safe manner.

These entities are obliged to submit an annual report on the state of the Dam and periodic inspections are undertaken for the main two large dams operated by the Power Company.

Small dams are under the authority of the regional and local authorities. Generally, these Dams are operated and maintained by the Water user's association. Truly these dams are not well managed and maintained.

The Ministry of Agriculture and water resources created Executing Agency Works Water and Rural Equipment (AGETEER). It is a state corporation board, and it is the prime construction manager for and on behalf of the State and its agencies, local authorities, associations and any public body or private projects and programs in the following main areas:

- hydraulic infrastructure and the development of rural areas including irrigation schemes, dams, wells and boreholes, hatchery;

- equipment and rural buildings;

- local development;

The AGETEER handles the following activities:

- studies of rehabilitation works and budgeting work;

- studies of implementation of new structures;

- control of work.

6.1.3. Dam safety supervision

The Ministry of Agriculture and water resources is responsible for the supervision of the Dam safety. This ministry has several departments dealing with these activities.

- The General Directorate for water resources is responsible to the delivery of Authorization for Construction after the evaluation of the design papers.

- The Bureau for Supervision and Control of Work and studies is responsible to monitor the construction of dams and also to evaluate the design documents for large dams.

- For the main large dams the World Bank guidelines are applied.

As stated above the government is preparing to install a Dam safety unit and all the required regulations and institutional framework.

6.2. TECHNICAL FRAMEWORK OVERVIEW

There are not any technical guidelines for the Dam Safety Regulations in Burkina Faso.

7.   CANADA

Reporter: Andy Zielinski

7.1.   MAIN PRINCIPLES OF DAM SAFETY MANAGEMENT

Canada is a commonwealth of 10 Provinces (drawing their power and authority directly from the Constitution Act) and 3 Territories (receiving their power and authority from the federal government) with their own governments and laws/acts and orders/decrees.

7.1.1.   Legal Framework for Dam Safety

- There is no central national Canadian act with respect to dam safety and constitutionally the Provinces are provided with the authority over "waters", except for international boundaries.

- Four Provinces have acts providing framework for dam safety regulations and the Canadian Dam Association (CDA) Dam Safety Guidelines provide supplementary guidance.

- Those Provinces which possess acts with respect to dam safety have more or less detailed orders and directives of the governmental administrations regarding dam safety.

- The Provinces and Territories without acts and regulations with respect to dam safety rely on Dam Safety Guidelines by Canadian Dam Association (see below).

The following table gives an overview of dam safety legislation in Canada:

State/Federal State/Province/Territory	Law/Act concerning dam safety	Decrees etc. concerning dam safety
Canada (Federation/central State)	no	no
Alberta	yes/d	yes
British Columbia	yes/d	yes
Ontario	yes/d	yes
Quebec	yes/d	yes
Newfoundland and Labrador	no	no
Nova Scotia	no	no
New Brunswick	no	no
Prince Edwards Island	no	no
Manitoba	no	no
Saskatchewan	no	no
Yukon	no	no
Nunavut	no	no
Northwest Territories	no	no
d = direct demands regarding dam safety		

Relevant documents of the Provinces are (selection):

- British Columbia Dam Safety Regulation, B.C. Reg. 44/2000, February 10, 2000.

- Alberta Regulation 205/98, Water Regulation – Part 6 Dam and Canal Safety.

- Quebec Dam Safety Act, R.S.Q. and following Orders in Council (2002 and 2005) regarding Dam Safety Regulation.

- Ontario Lakes and Rivers Improvement Act. R.S.O. (1990), Ontario Regulation 454/96 (1996).

- Series of Technical Bulletins and directives in each of these Provinces which provide guidance to owners and practitioners.

7.1.2. Responsibilities for Dam Safety

The dam owners are responsible for dam safety.

7.1.3. Arrangements for Independent Dam Safety Supervision

Provincial regulators have different powers and administrative arrangements to supervise the application of the dam safety regulations.

7.2. DAM CLASSIFICATION

7.2.1. General Principles of Classification

- Life safety (loss of life or population at risk), property (direct economic loss to third parties), environmental impacts, other socio-economic impacts and cultural heritage loss are criteria for classification of dams.

- Number of classes depends on potential incremental consequences of failure (4 to 6 classes varying between the Provinces).

7.2.2. Dam Classes Overview

The dam categories in the CDA Dam Safety Guidelines will be presented as example for the Canadian system of dam classification with the five classes as follows:

(Continued)

Dam class	Population at risk (PAR) [Note 1]	Incremental Losses		
		Loss of life [Note 2]	Environmental and cultural values	Infrastructure and economics
Low	None	0	Minimal short-term loss No long-term loss	Low economic losses, area contains limited infrastructure or services
Significant	Temporary only	Unspecified	No significant loss or deterioration of fish or wildlife habitat Loss of marginal habitat only Restoration or compensation in kind highly possible	Losses to recreational facilities, seasonal workplaces, and infrequently used transportation routes
High	Permanent	10 or fewer	Significant loss or deterioration of important fish or wildlife habitat Restoration or compensation in kind highly possible	High economic losses affecting infrastructure, public transportation, and commercial facilities
Very high	Permanent	100 or fewer	Significant loss or deterioration of critical fish or wildlife habitat Restoration or compensation in kind is possible but impractical	Very high economic losses affecting important infrastructure or services (e.g. highway, industrial facility, storage facilities for dangerous substances)
Extreme	Permanent	More than 100	Major loss of critical fish or wildlife habitat Restoration or compensation in kind is impossible	Extreme losses affecting critical infrastructure or services (e.g. hospital, major industrial complex, major storage facilities for dangerous substances)

Note 1. Definitions for population at risk:

None - There is no identifiable population at risk, so there is no possibility of loss of life other through unforeseeable misadventure.

Temporary - People are only temporarily in the dam-breach inundation zone (e.g., seasonal cottage use, passing through on transportation routes, participating in recreational activities.

Permanent - The population at risk is ordinarily located in the dam-breach inundation zone (e.g., as permanent residents): three consequence classes (high, very high, extreme) are proposed to allow for more detailed estimates of potential loss of life (to assist decision-making if the appropriate analysis is carried out).

Note 2. Implications for loss of life:

Unspecified - The appropriate level of safety required at a dam where people are temporarily at risk depends on the number of people, the exposure time, the nature of their activity, and other conditions. A higher class could be appropriate, depending on the requirements. However, the design flood requirements, for example, might not be higher if the temporary population is not likely to be present during the flood season.

In the case of six classes (only Province of Quebec), the consequence categories are as follows:

- very low,

- low,

- moderate/significant,

- high,

- very high, and

- severe.

7.3.   TECHNICAL FRAMEWORK OVERVIEW

There are central national Dam Safety Guidelines edited by the Canadian Dam Association (CDA) and several guidelines published by provincial governments.

Guidelines of Canadian Dam Association and some provincial governments (selection):

- Dam Safety Guidelines. Canadian Dam Association, 2007.

- Dam Safety Guidelines – Technical Bulletins. Canadian Dam Association, 2007.

- Guidelines for Public Safety around Dams. Canadian Dam Association, 2011.

- Dam and Canal Safety Guidelines. Publication No. T/444. Alberta Environment, March 1999.

- Technical standards and criteria for approval under the Lakes and Rivers Improvement Act. Ontario Ministry of Natural Resources, 2011 in form of Technical Bulletins.

- Dam Safety Guidelines – Inspection & Maintenance of Dams. BC Environment, 1998.

- BC Dam Safety Guidelines – Plan Submission Requirements for the Construction and Rehabilitation of Dams, BC Environment, 2008

7.4.   SUPPLEMENTARY REFERENCES

More information about dam safety management, legislation and classification can be found in the following references.

[1] Campbell, P. et al. Regulation of Dams and Tailings Dams in Canada. Proceedings of CDA 2010 Annual Conference. Niagara Falls, ON, Canada, October 2–7, 2010.

[2] www.damsafety.org/media/documents/STATE_INFO/STATISTICS/2008stateprogramstats.pdf

8.   CHINA

Reporter: Xu Zeping

8.1.   MAIN PRINCIPLES OF DAM SAFETY MANAGEMENT

The People's Republic of China exercises jurisdiction over 22 provinces, five autonomous regions, four directly administered municipalities (Beijing, Tianjin, Shanghai, and Chongqing), and two highly autonomous special administrative regions (SARs) – Hong Kong and Macau.

Laws and administration regulations (legislated by central government and local government) regulate dam safety.

8.1.1.   Legal framework for dam safety

The framework of the legal system concerning dam safety is basically composed by the following regulations.

- Regulations on the Registration of Reservoirs and Dams

- Regulations on the Registration of the Dams of Hydropower Station

- Regulations on the Safety Assessment of Reservoirs and Dams

- Detailed Rules for Implementation of Dam Safety Supervision of the Dams of Hydropower Station

- Administration Provisions for Reservoir Demolition and Decommission

The main laws and regulations concerning dams are:

- Water Law of the People's Republic of China (enacted in 1988 by State Council and revised in 2002)

- Flood Control Law of the People's Republic of China (published on Aug. 29, 1997, enacted at Jan. 1, 1998 by State Council)

- Managing Regulations on the Safety of Reservoirs and Dams (enacted at Mar. 22, 1991 by State Council)

- Managing Regulations on Water Conservancy and Hydropower Engineering (enacted at Apr. 23, 1983 by Ministry of Water Resources and Electric Power)

The main laws and regulations concerning construction the quality of dams are:

- Construction Law (Published on Nov. 1, 1997 and enacted on Mar. 1, 1998 by State Council)

- Managing Regulations on the Quality of Construction Projects (Enacted on Jan. 2000 by State Council)

- Managing Regulations on the Quality of Water Projects (Enacted on Dec. 21, 1997 by State Council)

- Provisional Regulations on the Treatment of Quality Accident of Water Projects (Enacted on Mar. 4, 1999 by Ministry of Water Resources)

Besides, according to the prescription of "Water Law of the People's Republic of China", "Flood Control Law of the People's Republic of China", etc. the water resources department can directly issue regulation orders to the administration office of the dam during emergency situations.

State/Country	Law/Act concerning dam safety	Decrees etc. concerning dam safety
China	yes	yes

8.1.2. Responsibilities for dam safety

The primary responsibility for dam safety lies with the owner of the dam. The project owner, designer, contractor and supervisor are responsible for the engineering quality of the project. The owner and the administration agency of the dam are responsible for the safe operation of the dam. Besides, there are also government legislations for prescribing the penalties on any actions that has impacts on dam safety, whether it is conducted by one´s private capacity, corporations or organizations.

The dam owner is obliged to have a surveillance program. The extent of operational dam safety measures, surveillance program etc. taken for each dam depends on the type and relevance of the dam. The dam safety supervision includes daily inspection, yearly detailed inspection, regular inspection and special inspection, etc. With the regular safety assessment and supervision, the safety properties and operation status of the dam can be known, and the defects and hidden dangers will be discovered. It is very helpful for making the decision for dam reinforcement and rehabilitation.

According to the regulations, the assessment on safety of reservoir impounding must be conducted before the first impoundment of the reservoir, when dam construction is completed. After the dam is put into operation, the first safety assessment will be conducted within 2~5 years after the first reservoir impoundment. For the dams in normal operation, the period for safety assessment is 6~10 years.

The dam operation program will also be approved in advance before the implementation.

8.1.3. Arrangements for independent dam safety supervision

According to the main functions of the project, the dams and reservoirs in China are managed by two different departments. Dams with their main functions for flood control, water supply and irrigation, are managed by government departments of water resources. Dams with their main functions for power generation are managed by the departments of electric power.

In the 1980s, the Ministry of Water Resources and the Ministry of Energy have set up Dam Safety Management Center and Large Dam Safety Supervision Center, respectively. Now, despite the Ministry of Energy and its successor State Power Cooperation was abrogated, the Large Dam Safety Supervision Center still take the responsibility for managing the dam safety of hydropower stations of the country.

Besides, according to the prescription of "Water Law of the People's Republic of China", "Flood Control Law of the People's Republic of China", etc. the water resources department can directly issue regulation orders to the administration office of the dam during emergency situations. For the existing defective and hazardous dams and reservoirs, the central government and local government will take measures to do the rehabilitation works gradually. Besides, it is also required by the Ministry of Water Resources to solve dam safety problems by demotion and decommission of the unsafe or disused dams according to the local conditions. For further regulating the works, the

Ministry of Water Resources has established a system for reservoir demotion and decommission and issued the "Administration Provisions for Reservoir Demotion and Decommission".

8.2. Technical Framework overview

The present standard system in China includes national standards, industry standards, local standards, enterprise standard, etc. The first three kinds of standards are compulsive.

8.3. Dam classification

Normally, dams are classified by considering dam height, reservoir storage capacity, failure consequence, etc. The classification relates to the functions of the dam and is prescribed in different technical standards.

9.   CZECH REPUBLIC

Reporter: Jiri Polacek

9.1.   MAIN PRINCIPLES OF DAM SAFETY MANAGEMENT

Czech Republic is a full democratic country managed by system of Upper and Lower House with central governance and president. The state legislation is unified. There are about 20 000 dams in Czech Republic, mostly small historical ponds, including 118 large dams according to the ICOLD criteria. The main purposes of large dams are hydropower production, drinking water, flood protection and water accumulation. Historical small dams serve mostly for fishing production.

9.1.1.   Legal framework for dam safety

- Water Act No. 254/2001 at valid version from 2010, two paragraphs No. 61 and 62 of it serve to dam safety only.

- Implementing regulation to water act No. 471/2001 at valid version from 2010 specifies all obligations of the dam safety surveillance and supervision.

State	Law/Act concerning dam safety	Decrees etc. concerning dam safety
Czech Republic	yes	yes

9.1.2.   Responsibilities for dam safety

In principle the dam owner is responsible for dam safety. The owner can be replaced by the dam user if the dam (usually small dam) is rented. The State owns most large dams. Five special state enterprises, divided according to main river basins, administer the function as dam owner.

According to the Water Act the owner has the obligation to cooperate with a company specialised on dam safety and surveillance authorised by the Agriculture Ministry on the basis of competition;

- For 1st and 2nd category dams the only responsible company has been VODNI DILA – TBD a/s for many years.

- For 3rd category dams there are now 11 cooperating companies. The authorised company regularly evaluates monitoring and measurement results and dam behaviour by means of expert reports and proposes some remedial measures.

- Only the owners are responsible for dams of the 4th category.

9.1.3.   Arrangements for independent dam safety supervision

According to regulation No. 471/2001 dam owners 1st, 2nd and 3rd category are obliged to have dam surveillance programme. Arrangements for Independent Dam Safety Supervision.

The supervision of dam safety is carried out in accordance with Water Act, § 62, by state or regional water authorities.

The Ministry of Agriculture check the dam surveillance performance at the highest categories I (every year) and II (each second year).

The Regional or Local (District) Water Authority check the dam surveillance performance at categories III (once four years) and IV (once ten years).

9.2.   TECHNICAL FRAMEWORK OVERVIEW

A large net of technical standards and guidelines are in function at the dam practice in Czech Republic. These documents are regularly upgraded.

Overview of main standards and guidelines connected with Dam Safety:

- Standard CSN 75 2310 Embankment dams.

- Standard CSN 75 2410 Small dams.

- Standard CSN 75 2935 Dam Safety assessment during floods.

- Standard TNV 75 2415 Dry reservoirs (polders).

- Standard TNV 75 2910 Manipulation orders of dams.

- Standard TNV 75 2920 Operational orders of dams.

- Standard TNV 75 2931 Flood plans.

- Guideline of Agriculture Ministry for Dam Safety Classification (categorization).

- Guideline of Agriculture Ministry for Dam Surveillance elaboration on dams of category IV.

- Guideline of Environment Ministry for securing of a warning and forecast service.

- Guideline of Environment Ministry for working-out of downstream area protection plan prior to flood effects caused by dam break.

9.3.   DAM CLASSIFICATION

9.3.1.   *General principles of classification*

The obligation of dam categorisation is anchored in § 61 of the Water Act.

Each constructor (owner) is obliged to give water-legal authorities a reference about the necessity and conditions of dam surveillance performance with a proposal of dam category. In case of a dam reconstruction the owner is obliged to review if the dam category is correct.

VODNI DILA – TBD is the only authorised company for elaboration of these references for all dam categories. Regional water-legal authority decides about final dam category on the basis of this reference.

According to regulation 471/2001 (valid version from 2010) the following hydraulic structures with impounded water (which will be released in case of structure failure) are divided in four classes marked by roman numbers I (the highest danger), II, III and IV:

a. dams and weirs,

b. structures for flood protection (dry reservoirs – polders, dikes – levees etc.),

c. structures of settling basins (tailings dams),

d. hydrotechnic tunnels or hydropower plants,

e. structures which are established for river navigation or in riverbeds or on their banks.

Dam classification in Czech Republic is established on the dam existence (basic parameters) and risk factor level of it in downstream area (range of human lives endangerment, damages on property and infrastructure, damages from losses of hydraulic structure utility and utilities of public interest). The risk factor is evaluated by counted number of points:

Category I	$F \geq 1,000$ points,
Category II	$150 \leq F < 1,000$,
Category III	$15 \leq F < 150$,
Category IV	$F < 15$ points

The value of one classification point is 2 millions of Czech crowns in 2012 (approximately 100 thousand USD).

CATEGORY	CRITERIA FOR RATING OF HYDRAULIC STRUCTURES
I	Thousands up to tens of thousands human lives losses are anticipated. Large damages of hydraulic structure the subsequent renewal of which is very complicated and expensive. Extensive damages to resident and industrial developments, road and railway system will arise and further hydraulic structures are endangered in the downstream area. Losses caused by the outage of hydraulic structure, by interruption of the industrial production, transports etc. are very high and difficultly reparable. Environmental damages are high, extending the importance of the higher administrative autonomous district, the economic consequences concern the whole country.
II	Hundreds up to thousands of human lives losses are anticipated. Essential damages of hydraulic structure, its subsequent renewal is complicated and expensive.
	Large damages to the resident and industrial objects, transport networks are expected downstream, endangered are other hydraulic structures. Losses caused by outage of hydraulic structure, interruption of industrial production, transports and other losses are considerable. Damages to the environment extend the importance of the higher administrative autonomous district.
III	Tens up to hundreds of human lives losses are anticipated. Damage of hydraulic structure, renewal is feasible. Damages downstream arise to the residential and industrial objects as well to the transport network, endangered could be further, less important hydraulic structures. The material damages in the downstream area are small. Losses caused by the outage of hydraulic structure are small. Damages to the environment are negligible.
IV	Human lives losses are improbable. Damage of hydraulic structure, renewal is feasible. Material damages downstream are small. Losses caused by the outage of hydraulic structure, by interruption of industrial production, transports and other losses are fully reparable.

9.4.   REFERENCES

[1] Act No. 254/2001 about water (Water Act) at valid version.

[2] Regulation No. 471/2001 about Dam Safety Surveillance at valid version.

[3] Regulation No. 590/2002 about technical demands for hydraulic structures at valid version.

[4] Regulation No. 195/2002 about requirements for manipulation and operational orders of dams.

10.   FINLAND

Reporter: Eija Isomäki

10.1.   MAIN PRINCIPLES OF DAM SAFETY MANAGEMENT

Finland is a republic with a central government. There are 15 Centres for Economic Development, Transport and the Environment (ELY Centres) in Finland. Supervision of dam safety has been centralized in three ELY Centres. Supreme supervision and guidance on dam safety issues belongs to the ministry of Agriculture and Forestry.

There are 435 dams that are classified according to the Dam Safety Act. 56 dams are large dams according to the ICOLD criterion. The number of dams has recently been increasing because of new mining projects.

10.1.1. Governmental arrangements concerning dam safety

The most important Act is the Dam Safety Act (494/2009) and Government Decree on Dam Safety (319/2010). Both are valid for all dams, including waste dams and tailing dams. The Dam Safety Act applies to all dams independent of the material of which the dam is constructed or the substance impounded by the dam. But there are only few requirements in the legislation that apply to non-classified dams.

There are some other acts and decrees cover the aspects of dam safety. For example:

* Water Act (587/2011)

* Government Decree on Water Management (1560/2011)

* Rescue Act (579/2011)

* Mining Law (621/2011).

State/County	Law/Act concerning dam safety	Decrees etc. concerning dam safety
Finland (all counties)	yes	yes

10.1.2. Responsibilities for dam safety

Dam Safety Act states that the owner of the dam has the task to see to the design, construction, operation and maintenance of the dam. So they have the total responsibility for their dams.

The owner of a dam, that has been classified, must organise a periodic inspection of the dam at least every five years. The dam safety authority and the rescue authority have the right to participate. During periodic inspections the dam owner must verify that the dam fulfils the safety requirements. If the periodic inspection reveals deviations from safety requirements, the owner must prepare a thorough study of the condition of the dam.

10.1.3. Arrangements for Independent Dam Safety Supervision

Since 1999 the governmental dam safety supervision has been centralized to three of the totally 15 regional centres "ELY Centres". These are located in the regions Häme, Kainuu and Lapland. They officially supervise all aspects of dam safety, except rescue procedures.

10.2. DAM CLASSIFICATION

10.2.1. General principles of classification

Based on the hazard, dams are placed in one of the three classes 1, 2 and 3. Dams are classified by the type of hazard they pose if an accident occurs to human lives, health, environment and property.

Classification is not needed (non-classified dam), if the dam safety authority considers that the dam does not cause any danger.

10.2.2. Dam classes overview

Class 1 dam, which in the event of an accident causes danger to human life and health or considerable danger to the environment or property
Class 2 dam, which in the event of an accident may cause danger to health or greater than minor danger to the environment or property
Class 3 dam, which in the event of an accident may cause only a minor danger.

If a dam does not cause any danger, it is denoted as a non-classified dam.

10.3. TECHNICAL FRAMEWORK OVERVIEW

The Dam Safety Guide (that explains and describes the legislation) was published in the internet in 2012 by ELY Centre for Häme. It will be translated into English and Swedish in 2013 (http://www.environment.fi/).

10.4. SUPPLEMENTARY INFORMATION

More information can be found in the internet: http://www.environment.fi/ > Water resources management > Dams and dam safety. The Dam Safety Guide will be published in this website.

11.   FRANCE

Reporter: Michel Poupart

11.1.   MAIN PRINCIPLES OF DAM SAFETY MANAGEMENT

France is a republic in Western Europe with several overseas territories and islands. The French Republic is defined as indivisible, secular, democratic and social by its constitution.

France is divided into 27 administrative regions (including Corsica and five overseas regions). The regions are further subdivided into 101 departments. The 101 departments are subdivided into 341 arrondissements which are subdivided into 4,051 cantons. These cantons are then divided into 36,697 communes - municipalities with an elected municipal council. The regions, departments and communes are all known as territorial collectivities, meaning they possess local assemblies as well as an executive. Arrondissements and cantons are merely administrative divisions.

The French parliament is a bicameral legislature comprising a National Assembly (Assemblée Nationale) and a Senate. The Senate's legislative powers are limited; in the event of disagreement between the two chambers, the National Assembly has the final say.

According to the ICOLD Register of dams there are 622 large dams in France, but 55 of them are lower or equally high than 15m. Over 250 of the existing dams have been built for hydroelectric production. First dams were built by Romans in Southern France in 1rst century B.C. mainly for water supply. Ever since many dams have been built, especially after year 1950, when the number of dams started to increase rapidly. In the last period dams are built mostly for power production and now for water supply and flood management.

11.1.1. Legal framework for dam safety

List of laws and decrees that concern dam safety:

*   Water license: A 1919 law gives the general rules that regulate the process of building and operation of dams for hydro electrical generation – precisely the use of water, the duration of the license, the requirement on water quality, environmental issues, and general obligations about public safety. This law has been followed by decrees last of them issued in 1994, 1999 and 2008.

*   Laws and decrees for industrial or public uses of water and water supply.

*   Law on Water and Aquatic Environment (2006) – establishes the "Technical Committee on Dam and Hydraulic Works" and defines its role and responsibilities; it states that a risk assessment is required for dams and dykes according to their classification.

*   A decree issued in December 2007 defines the dams and dyke's classification and tasks that should be fulfilled by owner or operator (design, studies, quality of construction, first filling, surveillance, monitoring, tests, reports to the authorities) according to their classification.

- Decrees issued in 2005 specify the requirements for emergency preparedness plans (EPP) for dams higher than 20 m above the natural ground level and with a reservoir capacity over 15 hm^3, or those which represent a high hazard for the population.

- Government order issued in 1999 which gives specific requirements to guarantee public safety around dams and waterways.

- Subsequent regulatory orders address the risk assessment methodology, the flood and seismic hazard requirements, the incident reporting, the consultants' technical agreements.

State/Country	Law/Act concerning dam safety	Decrees etc. concerning dam safety
France	yes	yes

11.1.2. Responsibilities for dam safety

The owner is the one responsible of dam safety. He has to perform operation, surveillance, maintenance of the dam, evaluate the consequences of dam failure, undertake periodical safety review and risk assessment. He prepares and sends periodical reports to the authority, and reports for incident or accident. He must implement an internal management system to supervise all the activities related to dam safety. The application of tasks required by law is controlled by Governmental regional regulatory bodies.

The requirements (risk assessment and safety review, existence of a monitoring system, periodicity of reports, historical files, etc.) for the dams vary according to the dam classification (see section 11.2 for classification definition):

The table below summarizes these requirements for dams:

Dam classification (A, B, C, D)	A	B	C	D
Characteristics and historic files	x	x	x	x
Monitoring system and visual inspection obligation	x	x	x	
Report about operation, surveillance, works,...	x	x	x	x
Periodic monitoring reports (behaviour)	x	x	x	
Risk assessment (1/10 year) (Etude de danger)	x	x		
Safety review (1/10 year)	x			
Management system	x	x	x	x
Incident reporting	x	x	x	x

11.1.3. Arrangements for independent dam safety supervision

The levels of control are performed at different levels:

```
┌──────────────────────────────────────────┐
│   Technical Committee on Dams             │
│       and Hydraulic Works                 │
│   approves designs, studies, EPP,         │
│   special remedial works                  │
└──────────────────────────────────────────┘
                    │
                    ▼
┌──────────────────────────────────────────┐
│      Governmental regional                │
│        regulatory bodies                  │
│   control the owner, carry out field's    │
│   inspection, issue recommendation        │
│   reports                                 │
└──────────────────────────────────────────┘
                    │
                    ▼
┌──────────────────────────────────────────┐
│          Dam owner                        │
│   surveillance, operation,                │
│   monitoring, maintenance                 │
│                                           │
│   Management system                       │
└──────────────────────────────────────────┘
```

At a national level, the Technical Committee on Dams and Hydraulic Works is responsible for the approval of new projects design, major rehabilitation works design, and EPP consequences assessment. Furthermore, an authorisation by the administration must be obtained before construction of dams higher than 2 m. For that authorisation, a risk assessment study ("étude de dangers") for dam classified in A and B category and dykes classified in A, B and C (see section 2.1 for classification definition) is required.

At the regional level Governmental regulatory bodies control that owners or licensees realize safety activities according to the law and decrees; they perform onsite inspection on a periodical basis, and control the technical reports, safety assessments and safety review as defined by the 2007 decree. They approve the owner procedures for surveillance, flood routing and normal operation.

If a dam or a dyke performance does not meet required criteria (either from standards limits or by expert judgment), these regional bodies require a "special review" of this dam or dyke. This special review includes a global design review, a risk assessment and proposals for remedial actions, either structural or non-structural. The regional regulator approves and prescribes the remedial actions or may ask for more detailed studies or for other actions. According to the dam classification and the hazard involved, the Technical Committee on Dams and Hydraulic works may be involved in this approval procedures.

A national specialized governmental body (about 10 engineers) supports the regional bodies for technical reviews of reports, procedures and risk assessment.

Finally, at the owner level, the management system defines organizational and technical framework, assigns responsibilities and tasks, specifies internal controls and periodical updates of the management system.

11.2. DAM CLASSIFICATION

The decree of 2007 defines the dam and dykes classification. This decree divides the dams into four categories:

Category	Criteria
A	All dams with: $H \geq 20$ m
B	Dams not in category A and with: $H > 10$ m and $H^2 * V^{1/2} \geq 200$
C	Dams not in category A or B and with: $H \geq 5$ m and $H^2 * V^{1/2} \geq 20$
D	Dams not in category A or B or C and with: $H \geq 2$ m
(H = height above natural ground level, V = reservoir volume in hm³)	

There is no classification considering potential risk of a dam directly. For the dams, only H and V are considered.

This classification is also used for the flood and earthquake hazard requirements.

For the dykes it is different since the number of people potentially affected by a dyke failure (potential damage) is taken into account for the classification ranking.

The decree divides the dykes into four categories:

Category	Criteria
A	Dykes with: $H \geq 1$ m and $P \geq 50000$
B	Dykes not in category A awith: $H \geq 1$ m and $1000 \leq P \leq 50000$
C	Dykes not in category A or B and with: $H \geq 1$ m and $10 \leq P \leq 1000$
D	Dykes with: $H < 1$ m and $P < 10$
(H = height above natural ground level, P = population potentially affected)	
Note: the dam category can be modified by authorities, in particular in case of important downstream consequences.	

11.3. TECHNICAL FRAMEWORK OVERVIEW

The technical framework consists of design guidelines, standards and technical recommendations.

Technical recommendations published by government exist for:

- Seismic assessment: hazard evaluation, seismic stability, acceptance criteria (2013),

- Spillway hydrological and hydraulic design (2013),

- Guidelines for regional regulators about all the aspects of dam safety,

- Recommendations for dam and dykes risk assessment (2008 and 2010).

Guidelines exist for:

- Probabilistic stability calculation for gravity dams (2012),

- Probabilistic stability calculation for earthfill dams (draft - 2010),

- Design and construction of small dams, Cemagref edition, Degoutte and al, 1997,

- Surveillance and maintenance of small dams, Cemagref edition, P. Royet, 2006.

Some owners with very important dam portfolio have developed their own more detailed guidelines and technical documents in full accordance with the regulation for internal use.

More information about dam safety management, legislation and classification can be found in [2] to [9] and on "http://www.legifrance.gouv.fr/" where all French laws and decrees are available.

11.4. REFERENCES

[1] http://en.wikipedia.org/wiki/France

[2] P. Le Delliou et al, New French regulations concerning dams, 24th ICOLD Congress, Q93 R14, 2012 Kyoto

[3] Cochet et al, Preparation of earthquake regulations for dams in France, 24th ICOLD Congress, Q93 R12, 2012 Kyoto

[4] Aigouy et al, Guidelines for design of dam spillways in France, 24th ICOLD Congress, Q94 R23, 2012 Kyoto

[5] Royet P., Peyras L., New French guidelines for structural safety of embankment dams in a semi probabilistic format, 8th ICOLD European club Symposium, 2010 Innsbruck

[6] Degoutte et al, New regulations on the safety of dams and levees in France, 23th ICOLD Congress, Q91 R40, 2009 Brasilia

[7] Bister D., Le Delliou P., Risk analysis and danger flood, 20th ICOLD Congress, Q76 R36, 2000 Beijing

[8] Bister et al, Dam safety, new approach of French regulations, 19th ICOLD Congress, Q75 R42, 1997 Durban

[9] Degoutte G. et al, New regulations on the safety of dams and levees in France, 23rd ICOLD Congress, Q91 R40, 2009 Brasilia

[10] http://www.legifrance.gouv.fr/ (official website with all laws, decrees, etc. in French)

12. GERMANY

Reporter: Hans-Ulrich Sieber

12.1. MAIN PRINCIPLES OF DAM SAFETY MANAGEMENT

Germany is a federal republic with 16 independent federal states with their own governments, laws, ordinances, decrees and administrative directives. Large dams exist in 12 of the 16 states.

In 2012, 312 German dams were in accordance with the criteria of the ICOLD World Register of Dams.

12.1.1. Legal framework for dam safety

The German legal framework for dam safety can be described as follows:

- Ensuring dam safety is the responsibility of the individual German states.

- There is no German federal law governing dam design or dam safety.

- The federal states with existing dams have Water Acts which don´t include explicit requirements or rules regarding dam safety. But they require that design, construction, operation and supervision of dams adhere to the current technical framework (e. g. DIN standards) for dams. These requirements are valid for dams with certain geometrical sizes defined by the water acts of the states. Small dams (usually those with height of dam < 5 m and storage capacity < 100,000 m³) are excluded from the legal requirements in the most cases.

- In addition, federal states with existing dams have additionally (different) statutory ordinances and/or administrative regulations on dams having effects on dam safety. These "lower" regulations also include small dams.

The following table gives an overview of dam safety legislation in Germany:

State/ Federal State/ Province/Territory	Law/Act concerning dam safety	Decrees etc. concerning dam safety
Germany (Federal Republic)	no	no
Baden-Württemberg	yes/i	yes
Bavaria	no	yes
Berlin	no	no
Brandenburg	yes/i	yes
Bremen	(yes)	no
Hamburg	no	no
Hesse	(yes)	yes
Lower Saxony	(yes)	yes
Mecklenburg-W. Pomerania	no	no
North Rhine-Westphalia	yes/i	yes
Rhineland-Palatinate	yes/i	yes
Saarland	(yes)	yes
Saxony	yes/i	yes
Saxony-Anhalt	(yes)	yes
Schleswig-Holstein	no	no
Thuringia	yes/i	yes
Rhineland-Palatinate	yes/i	yes

Note:

(yes) = only specific demands at permit issue

d = direct requirements regarding dam safety;

i = indirect requirements regarding dam safety (for instance by requiring state-of-the-art technology observing certain regulations);

12.1.2. Responsibilities for Dam Safety

Dam owners or operators are responsible for the safety of the dam throughout its lifecycle (from design and operation phase to reconstruction or decommissioning of the dam). Dam owners may be private persons or companies, municipalities, associations, states or the federal Water and Shipping Authority (only two dams).

12.1.3. Arrangements for Independent Dam Safety Supervision

All dam owners or operators are responsible for the self-supervision of their dams. Administrative forces (authorities) of the federal states supervise the activities of the dam owners or operators. The four-eyes-principle (double-verification principle) is ensured for the supervision of dams independent of their size. This principle is valid for dams in state ownership too. However, in practice the intensity of the monitoring depends on the size and purposes of reservoirs as well as risks and other criteria.

Dam owners are obliged to prepare an "annual dam safety report" for each dam. In addition, they have to perform a so called "deep investigation" for each dam every 10 to 15 years depending on the dam category. The annual dam safety report deals mainly with the results of the monitoring of the dam obtained by visual observations and measurements of its behaviour. The deep investigation includes additional reviews of all design criteria and assumptions for dam safety calculations. The resulting reports are to be submitted to the responsible supervisory boards. They review them and if necessary, request measures of the dam owner or operator.

12.2. DAM CLASSIFICATION

12.2.1. General Principles of Classification

- Height of dams and volume of reservoirs are criteria for their classification.

- Significance of dams and hazard potential connected with dams are additional criteria to be considered qualitatively.

- Number of classes (categories) varies by the types of reservoirs (dams, flood retaining reservoirs, tailings dams, weirs); two to four classes (categories) are intended.

12.2.2. Dam Classes Overview

Dams of permanently impounded water storage reservoirs (DIN 19700-11):

- Category 1 – large dams: height > 15 m or gross storage capacity > 1,000,000 m³

- Category 2 – medium and small dams: all dams not included in category 1

The most important design criteria depending on dam classes are design floods and design earthquakes. The following table gives an overview of the sizes to be applied.

Dam category	Floods		Earthquakes	
	design flood	safety flood	operational earthquake	design (safety) earthquake
1 (large dams)	1,000-year-flood	10,000-year-flood	500-year-earthquake	2,500-year-earthquake
2 (small ... medium dams)	100...500-year-flood	1,000...5,000-year-flood	100-year-earthquake	1,000-year-earthquake

Dams of temporarily impounded flood retaining reservoirs (DIN 19700-12):

- Category 1 – large: height > 15 m or gross capacity > 1,000,000 m³

- Category 2 – medium: not included in category 1 and height > 6 m or gross capacity > 100,000 m³

- Category 3 – small: not in category 1 or 2 and height > 4 m or gross capacity > 50,000 m³

- Category 4 – very small: not included in category 1, 2 or 3

12.3. TECHNICAL FRAMEWORK OVERVIEW

There are federal technical guidelines concerning dams and dam safety. They consist of the national standards of the German Institute of Standardization (DIN) and several guidelines of scientific – technical associations (especially German Association for Water, Sewage and Waste; DWA). National Standards (selection of the most important for dams):

- DIN 19700-10 Stauanlagen: Gemeinsame Festlegungen, 2004 (Dam plants - Part 10: General specifications)

- DIN 19700-11 Stauanlagen: Talsperren, 2004 (Dam plants - Part 11: Dams)

- DIN 19700-12 Stauanlagen: Hochwasserrückhaltebecken, 2004 (Dam plants - Part 12: Flood retaining basins)

- DIN 19700-13 Stauanlagen: Staustufen, 2004 (Dam plants – Part 13: Weirs)

- DIN 19700-14 Stauanlagen: Pumpspeicherbecken, 2004 (Dam plants – Part 14: Pump storage basins)

- DIN 19700-15 Stauanlagen: Sedimentationsbecken, 2004 (Dam plants – Part 15: Tailings ponds)

- DIN 19702 Standsicherheit von Massivbauwerken im Wasserbau, 2010 (Stability of concrete and masonry hydraulic structures)

- DIN 1054 Baugrund: Sicherheitsnachweise im Erd- und Grundbau – Ergänzende Regelungen zu DIN EN 1997-1, 2010 (Subsoil: Verification of safety of earthworks and foundations – supplementary rules to DIN EN 1997-1, 2010)

- DIN EN 1997-1 Eurocode 7: Entwurf, Berechnung und Bemessung in der Geotechnik – Teil 1: Allgemeine Regeln, 2009 (Eurocode 7: Geotechnical design – Part 1: General rules, 2009)

Guidelines of DWA (formerly DVWK) (selection):

- DVWK Merkblatt M 231/1995: Sicherheitsbericht Talsperren – Leitfaden (Safety Reports – Guideline)

- DVWK Merkblatt M 246/1997: Freibordbemessung an Stauanlagen (Freeboard allowance of dams)

- ATV-DVWK Merkblatt M 502/2002: Berechnungsverfahren für Staudämme - Wechselwirkung zwischen Bauwerk und Baugrund (Calculation methods of fill dams – interaction between structure and foundation)

- DWA-Merkblatt M 507-1/2011: Deiche an Fließgewässern (River dykes)

- DWA-Merkblatt M 514/2011: Bauwerksüberwachung an Talsperren (Supervision of dams)

- DWA-Merkblatt M 522/2013 (Draft): Kleine Talsperren und Hochwasserrückhaltebecken (Small dams and reservoirs)

12.4. SUPPLEMENTARY REFERENCES

More information about dam safety management, legislation and classification can be found in the following references.

[1] Sieber, H.-U.; Hazard and risk assessment considerations in German Standards for dams – present situation and suggestions, Q. 76, R. 43, ICOLD Congress Beijing (2000).

[2] Rissler, P.; Talsperrenpraxis (Practice of dams), published by R. Oldenbourg Verlag Munich, Vienna, 1998.

[3] Dams in Germany; German Committee on Large Dams, published by Verlag Glückauf GmbH, Essen (2000).

13.   GREAT BRITAIN

Reporter: Andy Hughes

13.1.   MAIN PRINCIPLES OF DAM SAFETY MANAGEMENT

Great Britain comprises the territory of England, Scotland and Wales. England, Scotland and Wales forms the United Kingdom along with Northern Ireland. UK is a constitutional monarchy and a unitary state. England and Wales are under English Law, while Scotland has its own Scots law. The dam safety authority of England and Wales is the Environment Agency. The following facts are valid for England and Wales only.

There are approximately 2,500 dams in England and Wales subject to legislation, whereas about 800 dams are large dams/dams higher than 15 m.

13.1.1. Legal framework for dam safety

Requirements regarding dam safety are given in the Reservoirs Act 1975, HMSO, 1975 and the Floods and Water Management Act, 2010 – passed through Parliament but yet to be commenced.

State/Country	Law/Act concerning dam safety	Decrees etc. concerning dam safety
Great Britain (England and Wales)	yes	yes

13.1.2. Responsibilities for dam safety

- The dam owners are entirely responsible for the dam safety.

- Enforcement only by Government (The Environment Agency).

- Inspecting Engineers and Supervising Engineers from a Panel of Engineers.

13.1.3. Arrangements for independent dam safety supervision

Any dam greater than 25,000m³ above the natural ground has to be inspected at least every 10 years. Panel Engineers inspect against guidelines and recommend works if required. Recommendations in the interests of safety have to be completed and can be enforced.

Condition is a standards-based approach. Supervision by a trained engineer involves a visit at least once a year.

The Floods & Water Management Act, 2010 will move towards a risk/consequence based approach.

13.2. DAM CLASSIFICATION

13.2.1. General principles of classification

Dams are classified according to consequence of failure. Criteria for classification: life safety (loss of life or population at risk) and economic losses. Dam classification is moving to consequence categorisation and risk rather than just on retained volume. Classification is equal for all dams/ reservoirs (not type dependent)

Number of classes: 4

13.2.2. Dam classes overview

Currently dams are classified via guidelines associated with the Reservoirs Act, 1975 according to consequence of failure.

A	10 or more killed
B	0–10 or more killed
C	no one killed but economic loss
D	no losses

Under the new legislation the consequence classes will change to high significance and low – these have yet to be designed.

13.3. TECHNICAL FRAMEWORK OVERVIEW

There are guidelines on floods and seismicity:

- Floods: An Engineering Guide & Reservoir Safety: ICE 1992.

- An Engineering Guide to Seismic Risk to Dams in the United Kingdom – BRE 1991.

13.4. SUPPLEMENTARY INFORMATION

More information about dam safety management, legislation and classification can be found on the BDS (British Dam Society) web site and numerous papers from international journals and conference proceedings.

[1] Reservoirs Act 1975 – HMSO

[2] Floods & Water Management Act, 2010, -HMSO

14. GREECE

Reporter: George Dounias

14.1. MAIN PRINCIPLES OF DAM SAFETY MANAGEMENT

Greece is a unitary state with 13 regions. Seven decentralized administrations group one to three regions for administrative purposes on a regional basis.

There are approximately 250 dams in Greece of which 154 are large dams according to the ICOLD criteria. Approximately 40 large dams are currently under construction. The main purposes of large dams are power production, irrigation and domestic water supply.

14.1.1. Legal Framework for Dam Safety

Dam Safety Regulations or Guidelines do not exist in Greece. However, a recent law concerning environmental licensing imposes the dam owner to provide a safety surveillance programme for all dams (both existing and new dams).

Also, legislation for dam design imposes a dam break study to be conducted in order to proceed with licensing for construction of a dam. In addition, an Emergency Plan has to be drawn up for all dams. This task is carried out by the regional administrations, where the dam is located, with the assistance of the dam owner.

State	Law/Act concerning dam safety	Decrees etc. concerning dam safety
Greece	yes	no

14.1.2. Responsibilities for Dam Safety

The owner of the dam is responsible for dam safety.

14.1.3. Arrangements for Independent Dam Safety Supervision

No state supervision of dam safety exists in Greece. However, a proposal for legislation introducing a Dam Safety Authority has been prepared by the Greek Committee on Large Dams (GCOLD). The proposal is currently under discussion.

14.2. DAM CLASSIFICATION

Dam classification exists only in Environmental legislation as follows (KYA 1958, 2012, FEK 21B, 13/1/2012):

- A1: H> 50m

- A2: 5m<H<50m

- B: H<5m (catchment area and environmental protection also contribute)

A dam classification is included within the proposed dam safety legislation (now under discussion).

14.3. TECHNICAL FRAMEWORK OVERVIEW

In the absence of Dam Safety Regulations and Guidelines each dam owner draws up and follows his own rules. No general rules for surveillance routines etc. related to dam categorization exist. Instead, operational rules depend on dam characteristics, conditions in the downstream area etc.

As an example, the dam safety organisation of the largest dam owner in Greece; the Public Power Corporation S.A. (PPC SA) owning the largest dams with a total reservoir capacity exceeding 80% of the whole water volume stored in all Greek reservoirs, is described below:

- A Dam Monitoring and Safety Sector is arranged within PPC SA/ Hydroelectric Generation Department (HGD) headquarters in Athens. The Sector comprises of personnel with qualifications in geotechnical engineering, geology, seismic risk etc.

- A dam surveillance team, consisting of at least two persons (technicians) is present at all major dams. Minor dams are surveyed by the teams based on the nearest large dam. Dam surveillance is the sole task of the team, i.e. they are not involved in other activities.

- The dam surveillance team is responsible for collecting data on a regular basis from various instruments installed, recording them and providing digital time series of the data. The frequency of measurements depends on the particularities of each dam etc.

- Data collected are forwarded to the Dam Monitoring and Safety Sector in Athens for elaboration and evaluation. A Report is prepared each year by the Sector for each dam on a routine basis and is submitted to the HGD Director.

- Personnel from the Dam Monitoring and Safety Sector visit all dams at regular time intervals, in order to verify compliance of monitoring with the specified standards.

- Extraordinary events, if any, are urgently evaluated and precautionary measures, if necessary, are taken jointly by the HGD Director and the Power Station Manager.

15.   ICELAND

Reporter: Bjorn Stefansson

15.1.   MAIN PRINCIPLES OF DAM SAFETY MANAGEMENT

Iceland is a parliamentary republic and unitary state. There is no dam safety authority in Iceland.

15.1.1. Legal framework for dam safety

There are no governmental arrangements and no legislation concerning dam safety.

State/Country	Law/Act concerning dam safety	Decrees etc. concerning dam safety
Iceland	no	no

15.1.2. Responsibilities for dam safety

The dam owners are responsible for the dam safety. Almost all dams are owned by public utilities. Dam owners are required to prepare emergency response plan for each dam (or dam cascade) in co-operation with the Emergency Response Units at governmental and local level.

15.2.   DAM CLASSIFICATION

15.2.1. General principles of classification

Iceland follows the same principles as in Norway (based on Norwegian regulations for classification from 2001), i.e. criteria for classification are; danger to lives, structural damages and environmental damages.

Number of classes: 4

15.2.2. Dam classes overview

There are four dam classes in accordance with consequences of failure as follows:

- No or very small risk to people and minimal environmental and structural damages

- Small risk to people and little environmental and structural damages

- Possible risk for people in 1–20 houses (or summerhouses).

- Possible risk for people in more than 20 houses (or summerhouses)

15.3. TECHNICAL FRAMEWORK OVERVIEW

There are no official Icelandic standards or guidelines for dams. Dam design has followed and follows generally Norwegian regulations. Recent Icelandic legislation (2010) requires all power plant design drawings to be submitted for review to the local building authority (at community level).

The Icelandic dam owners use guidelines from the US Bureau of Reclamation (BUREC) for dam monitoring. Monitoring reports are issued annually on each dam by most dam owners. The owner of most Icelandic dams (Landsvirkjun) has a rigorous monitoring system as part of the quality management system of the company.

16.   INDIA

Reporter: Sunil Sharma

16.1.   MAIN PRINCIPLES OF DAM SAFETY MANAGEMENT

India is a federation composed of 28 states and 7 union territories. All states, as well as two of the union territories have their own governments and legislation. The remaining five union territories are directly ruled by the Centre. There are about 5100 dams in India, including 3335 large dams higher than 15 m according to the ICOLD criteria. The main purposes of large dams are irrigation, water supply and hydropower production.

16.1.1. Legal Framework for Dam Safety

In 2010 the Government of India, Ministry of Water Resources, formulated a Dam Safety Bill. The proposed Dam Safety Legislation will provide for proper surveillance, inspection, operation and maintenance of all dams of certain parameters (so called "specified dams") to ensure their safe functioning. At first instance the proposed Act, after the enactment, will not extent to all of India but to two states (Andhra Pradesh and West Bengal) and the Union Territories. To apply also to other States the legislation will have to pass their respective legislative arrangements.

Some states have enacted their own dam safety legislations. For example, the Government of Bihar has passed the "Dam Safety Act 2006" and the Government of Kerala has passed "The Kerala Irrigation and Water Conservation Act 2003" with some amendments in 2006 pertaining to dam safety. Currently the states of Madhya Pradech, Maharashtra and Uttar Pradesh are actively considering formulation of their own laws of dam safety.

The following table gives an overview of dam safety legislation in India:

State/Territory	Law/Act concerning dam safety	Decrees etc. concerning dam safety
India (Central State)	no (proposal presented 2010)	no
States and Union Territories	yes/no – varying	no

16.1.2. Responsibilities for Dam Safety

Dam safety is the responsibility of the dam owner. In India most large storage dams are owned, constructed and maintained by the State Governments or their Public Sector Undertakings and the Central Government Undertakings. Only a few dams are owned and operated by private bodies. Thus, the safety of the dams is the principal concern of the State Agencies/Organizations that own the dams and are involved in investigations, planning, design, construction, operation and maintenance.

16.1.3. Arrangements for Independent Dam Safety Supervision

As the practices of dam safety can vary from State to State, the Centre had been working towards evolving unified practices of dam safety and has recommended its implementation by all States/organisations. In 1979 a Dam Safety Organization (DSO) was established in the Central Water Commission (CWC) by the Government of India. DSO has made great efforts in creating awareness in the country and has succeeded to a large extent in convincing the States towards the concept of dam safety which has now been accepted by a large number of States. DSO has published several guidelines for dam safety and assisted the State Governments in deficiency investigations on dams and suggested remedial measures. Some of the States with many dams (Andhra Pradesh, Bihar, Gujarat, Kerala, Tamil Nadu, Maharashtra, Madhya Pradesh, Orissa, Karnataka, Rajasthan and Uttar Pradesh) have also created their own DSOs with varying capabilities.

In 1979 the Government of India also constituted the National Committee on Dam Safety (NCDS) in CWC. Almost all States with many dams are represented in the Committee. The Committee oversees dam safety activities in various States/Organizations and suggests improvements to bring dam safety practices in line with the latest state-of-art technology consistent with Indian conditions, and acts as a forum for exchange of views on remediation techniques. In 1991 also a National Committee on Seismic Design Parameters was constituted in CWC, and guidelines have been established.

Only in the States and Union Territories where dam safety legislation is in place dam owners are obliged to have a dam safety surveillance program for each dam. With the proposed Central Dam Safety Bill 2010, the dam safety surveillance program may become mandatory for a large number of dam owners.

It is general practice that all State agencies/corporations who own dams carry out the surveillance and maintenance on their own. The practices vary from State to State and agency to agency within the State. Generally, the owners' ordinary and special event driven inspections are carried out according to guidelines by DSO of CWC. In case of serious problems, the DSO's carry out further investigations and suggest remedial measures. Some States have constituted independent panels of experts, Dam Safety Review Panels, comprising of retired dam safety experts. Some owners conduct comprehensive dam safety reviews through the Dam Safety Review Panels.

A copy of the dam safety reports is generally sent to the DSO in CWC by the respective DSOs of the State Governments/dam owning organizations.

16.2. DAM CLASSIFICATION

Dam classification based on hazard potential has not been implemented by any State. However, the proposed Dam Safety Bill 2010 includes vulnerability classification of dams. Presently dam safety measures are mostly implemented on the basis of the distress condition of the dam and maintenance requirements, without regard of the potential failure consequences.

For the limited purpose of determination of design flood dams are classified according to gross storage capacity and hydraulic head.

Design flood classification:

Classification	Gross storage, million m³	Hydraulic head, (m)
Small	0.5–10	7.5–12
Intermediate	10–60	12–30
Large	> 60	> 30

16.3. Technical Framework Overview

CWC has prseveralmber of guidelines on dam safety and circulated these for adoption to the State Governments and other organizations associated with dam safety:

- Report on Dam Safety Procedures (1986)

- Guidelines for Safety Inspection of Dams (1987)

- Guidelines for Development and Implementation of Emergency Action Plan for Dams (2006)

- Standardised Data Book Format, Sample Checklist of Proforma for Periodical Inspection of Dams

16.4. SUPPLEMENTARY INFORMATION

Government of India, with World Bank assistance, is in the process of implementing the Dam Rehabilitation and Improvement Project including rehabilitation of 223 dams in four States and strengthening the dam safety institutional set up of the country. As part of the projects the existing guidelines will be updated, and new guidelines will be formulated.

Guidelines are available on the CWC website: http://cwc.gov.in/Dam_Safety.html

17.   INDONESIA

Reporter: Abdul Hanan Akhmad

17.1.   MAIN PRINCIPLES OF DAM SAFETY MANAGEMENT

Indonesia, which consists of 33 provinces, is an archipelago country with 5 large isles: Java, Sumatra, Kalimantan, Sulawesi and Papua.

Indonesia has 284 dams where 257 of them are state-owned and 27 dams do not belong to the government. All of them are classified.

17.1.1. Legal framework for dam safety

Indonesia has the Act 7 of 2004 on Water Resources that regulates control of destructive force of water, one of them is relating to reservoirs/dams.

Following up on the mandate of Act 7 of 2004, the government issued Government Regulation 37 of 2010 which regulates various aspects related to the construction of dams, such as, management, safety, finance, and control and the role of society.

Before the publication of the Act and Regulation, the Ministry of Public Works has issued rules regarding the safety of dams in Ministerial Decree No.378 in 1987 on Guidelines for Dam Safety and Ministerial Decree No.72 of 1997 on Safety of Dams.

State/ Federal State/ Province/Territory	Law/Act concerning dam safety	Decrees etc. concerning dam safety
Indonesia	Yes	Yes

17.1.2. Responsibilities for dam safety

There are three agencies responsible for the implementation of dam safety:

Technical Agency Safety of Dams; in this case is the Ministry of Public Works is in charge and shall:

- conduct an assessment of dam safety evaluation;

- provide recommendations on the safety of dams, and

- conduct dam inspections.

Technical unit area of dam safety; in this case is the Commission on Dams Safety supported by Balai Bendungan shall:

- provide technical support to the technical dam safety agency

Dam Owners and Dam Operator are in charge of:

- dam safety evaluation, monitoring and checking the condition of the dam.

17.1.3. *Arrangements for independent dam safety supervision*

Supervision of the dam conducted by the Commission on Dams Safety, whose members are government agencies, dam owners, representatives of professional associations in the field of dam and related Government agencies. In performing its duties, the Commission of Dams Safety is supported by Balai Bendungan, examples of task are:

- Data collection and processing as well as related programming

- Dam safety assessment for approval

- Periodic inspections and exceptional

- Implementation of dam behavior analysis

- Preparation of dam technical advice

- The implementation of cooperation with relevant agencies and the owners of the dam

- Socialising / publishing and mentoring about dam

- Preparation of regulations, guidelines, technical guidelines dam

- Inventory, registration and dam hazard classification

- Implementation and administrative affairs.

17.2. DAM CLASSIFICATION

Dams are classified as follows:

- dam with height of fifteen (15) meter or more measured from the bottom of the deepest foundation;

- dam with height of ten (10) to 15 (fifteen) meter measured from the deepest foundation of the following conditions:

    - dam crest length of at least 500 (five hundred) meter;

    - reservoir capacity of at least 500,000 (five hundred thousand) cubic meter, or

    - the maximum flood discharge calculated at least 1,000 (one thousand) cubic meters per second, or

- dam with difficult foundation or dam designed using new technology and/or dam having a high hazard class.

In Indonesia the dam safety level is classified into four categories:

1	Low danger level
2	Medium danger level
3	High danger level
4	Very high level of danger

The division of the hazard levels is based on the number of families at risk when exposed to the risk of the dam collapse with the assumption that each family consists of five people and live in a house. The division of the hazard rate is as follows:

Number of Families	Distance from Dam (km)				
	0–5	0–10	0–20	0–30	0 – > 30
0	1	1	1	1	1
1–20	3	3	2	2	2
21–200	4	4	4	3	3
> 200	4	4	4	4	4

17.3. TECHNICAL FRAMEWORK OVERVIEW

Indonesia has a number of security-related guidelines and guide of dam that has been released from 1998 issued by the Dam Safety Technical Agency, Ministry of Public Works. Until 2011, there have been 28 guidelines and 5 guidance associated to Dam.

List of Dam guidelines in Indonesia:

- Guidelines for Determining Dam Hazard Classification (1998)

- Guidelines for Preparation of Emergency Action Plan (1998)

- Code of Safety of Dams Commission Meeting (2002)

- Approval Procedures for the Development and Function Dam Removal (2002)

- Manual Charging Reservoir (2002)

- Guidelines for Inspection and Evaluation of Safety of Dams (2003)

- Guidelines for Dam Safety Studies (2003)

- General Design Criteria Guidelines for Dams (2003)

- Manual Operation, Maintenance and Dam Observation Part 1: General (2003)

- Manual Operation, Maintenance and Dam Observation Part 2: Operation and Maintenance Management (2003)

- Manual Operation, Maintenance and Parts Dam Observation 3: Instrumentation and Monitoring System (2003)

- Manual Operation, Maintenance and Parts Dam Observation 4: Security Inspection Equipment Hydromechanical (2003)

- Manual Operation, Maintenance and Dam Observation Part 5: Operations & Maintenance Hydromechanical & Electrical Equipment (2003)

- Guidelines for Management of Reservoir Sedimentation (2004)

- Guidelines for Mine Waste Dams (2004)

- Guidelines for Earthfill Dam Construction (2004)

- Visual Inspection Manual Earthfill Dam (2004)

- Seepage Control Guidelines on Earthfill Dams (2005)

- Guideline Grouting for Dams (2005)

- Cut-Off Wall Preparation guidelines (2005)

- Guidelines Earthfill dam on Soft Soil Foundation (2007)

- Guidelines for Dynamic Analysis of Earthfill Dam (2008)

- Guidelines for Dynamic Analysis of Concrete Gravity Dam (2009)

- Guidelines for Surveys and Monitoring Reservoir Sedimentation (2009)

- Technical Guidance for Dam Hazard Classification (2011)

- Technical Guidelines for Planning and Implementation of Solid Rolled Concrete Dams (2011)

- Technical Guidelines for Design and Construction of Rockfill Concrete Membrane Dams (2011)

- Dams Risk Assessment Technical Guidelines (2011)

List of Dam Guides in Indonesia:

- Earthfill Dams Planning Guide Volume 1: Survey and Investigation (1999)

- Earthfill Dams Planning Guide Volume 2 : Hydrological analysis (1999)

- Earthfill Dams Planning Guide Volume 3 : Foundation and Core Design (1999)

- Earthfill Dams Planning Guide Volume 4 : Complementary Building Design (1999)

- Earthfill Dams Planning Guide Volume 5 : Work Hydro Mechanical, Instrumentation and Building Supplement (1999)

17.4. SUPPLEMENTARY INFORMATION

More information on dam safety management, legislation and classification can be found on the webpage of Ministry of Public Works, www.pu.go.id and Indonesia National Committee on Large Dams (INACOLD), www.knibb-inacold.com.

18.   IRAN

Reporter: Mohsen Ghaemian

18.1.   MAIN PRINCIPLES OF DAM SAFETY MANAGEMENT

Regulations and orders of dam safety in Islamic Republic of Iran are organized by the Ministry of Energy. Regulations and orders consist of inspection and monitoring orders for dams under operation. The Ministry also organizes committees of "safety and stability of dams under construction and operation". A list of dams eligible for inspection and monitoring is published by ministry of energy every year.

Some orders issued in this manner are as follows:

List of dams eligible for inspection and monitoring, 2011–2012	Office of Water Supply Facilities Operation	6 July 2011
Necessity of monitoring and operation reports immediately after impounding	Iranian Water Resources Management Company	7 November 2010
Framework, tasks and responsibilities of "safety and stability of dams under construction and operation committees"	Iranian Water Resources Management Company	1 November 2009

18.1.1. Legal framework for dam safety

State/ Country	Law/Act concerning dam safety	Decrees etc. concerning dam safety
Iran	Yes	Yes

18.1.2. Responsibilities for dam safety

Responsibility of dam safety is not assigned explicitly in regulations however the owner of the dam is responsible for dam safety according to the public laws in Islamic Republic of Iran.

For example, according to the regulations issued by the ministry of energy, the owner of the dam must have dam safety surveillance program.

18.1.3. Arrangements for independent dam safety supervision

Governmental supervision of dam safety consists of assessment of inspection and monitoring reports, inspection of dam and facilities, ordering the owner to take any measurements and to conduct appropriate dam safety program.

18.2.   DAM CLASSIFICATION

18.2.1. General principles of classification

Dam classifications are based on the criteria of the regulations. Number of classes is different based on the subject of the classification. For dam safety program, ignoring some special cases, dams are classified to three categories considering their height and volume of the reservoir. Water-filled reservoirs and flood protection reservoirs are classified similarly.

Number of classes: 3.

18.3. TECHNICAL FRAMEWORK OVERVIEW

There are dam safety guidelines and standards in Islamic Republic of Iran. They have been prepared by professional experts under supervision of ministry of energy.

The main frameworks are as follows:

Guidelines for dam monitoring (200)	Office of Engineering and Technical criteria for water and waste water	1997
Guidelines for inspection of large dams (216)	Office of Engineering and Technical criteria for water and waste water	2001
Monitoring of underground excavations (252)	Office of Engineering and Technical criteria for water and waste water	2005
Guidelines for operation and maintenance of hydroelectric power plant in large dams (316)	Office of Engineering and Technical criteria for water and waste water	2010
Guidelines for seismic network design in dams (349)	Office of Engineering and Technical criteria for water and waste water	2010
Guidelines for Safety Assessment and Emergency Action Plan for Dams and Appurtenant Structures (370)	Office of Engineering and Technical criteria for water and waste water	2010
Criteria for selection of design Flood in Iranian large dams (378)	Office of Water Supply Facilities Operation	2007

19.   ITALY

Reporters: Carlo Ricciardi, Giovanni Ruggeri

19.1.   MAIN PRINCIPLES OF DAM SAFETY MANAGEMENT

Italy, officially the Italian Republic, is a unitary parliamentary republic subdivided into 20 regions (*regioni*), five of these regions have a special autonomous status that enables them to enact legislation on some of their local matters. The country is further divided into 110 provinces (province) and 8,100 municipalities (*comuni*).

The number of large dams in Italy is 542. 519 of these are higher than 15 m (Register of dams 2012). The number of dams has been increasing during the last decades, following the demand of electricity (hydro-power plants), irrigation, flood and debris control and water supply.

The dams are divided in two basic groups (large dams and small dams) – the safety of dams in each group is organized by different "authority":

- Large dams - the supervision is performed by National Dams Authority

- Small dams - are supervised by Regional Administrations

19.1.1. Legal framework for dam safety

There are several laws and decrees that concern dam safety. The law that provides the basis for the safety management of dams is Water and hydroelectric power plants Consolidation Act ("*Testo unico delle disposizioni di legge sulle acque e sugli impianti elettrici*", R.D. n° 1775). The law regulates the use of surface and underground water, including artificial reservoirs and with a particular attention to dams with the hydroelectric purpose.

Regulation that regulates dam safety more in detail:

- The Regulation for the Design, Construction and operation ("*Regolamento per i progetti, la costruzione e l'esercizio delle dighe di ritenuta - Parte I DPR n° 1363*") defines the general and administrative rules for the design, construction and operation of dams including the specification of requirements for different design level, the approval and authorization process, construction phase, supervision activity of the Authority, first filling and the final test for starting the normal operation, safety control during the operation stage.

- Urgent Measures concerning Dams law ("*Misure urgenti in materia di dighe*", Law n°584.) updates the definition of the "large dams" subjected to national legislation and authority. It also defines the procedure to be followed for the regularization of dams that were put in operation without having fulfilled the complete authorization process required by the Regulation. The technical and non-technical documents requested for the regularization are defined, as well as the measures to be taken by the authority if the regularization procedure is not fulfilled. It introduces the obligation for the owner to appoint for each dam an engineer who is responsible for the regular operation and safety of the dam.

- Regulation for the organization, duties and activities of the Italian Dam Authority ("*Regolamento concernente l'organizzazione, i compiti ed il funzionamento del*

Registro Italiano Dighe RID", DPR n° 136.) - The national authority is within the organization of the Ministry of infrastructure and transports. The decree defines the Italian Dam Authority organization. The regulation introduced the "Council of the dam owners" which should be consulted about subjects of main interest for the dam owners.

- Operational Directions pertaining to Dams (*"Disposizioni attuative ed integrative in materia di dighe", Circular letter n° DSTN/2/22806.*) - This Circular gives operative directions about appointment of the engineer responsible for the safety and its regular operation of the dam, detailed technical directions for the flood propagation studies, obligation of a half-yearly asseveration, about the safe conditions and operation of the dam, clarifications about the modalities for the evaluation of the basic dimensional parameters (dam height and reservoir volumes).

- Law n.214, 22.120.2011. The Ministry of infrastructures and transports (according to the proposal of Dam Authority), within 31-12-2012 has to release the list of the existing dams to improve for updated environmental loads according to the seismic and hydrological new knowledge. The Law requires also measures for removal of sediment, when bottom outlet obstruction may occur, and operational sedimentation control measures according to downstream environmental requirements.

Regulations and acts that influence directly or indirectly the dam safety:

- Financial Act for the year 2007 (*"Legge Finanaziaria 2007" Law n. 286, 24.11.2006*). The act cancelled the RID (National Dam Authority) as an autonomous body and returned it within the Ministry of the Public Works.

- Directions for Civil Protection activities in basins where dams are present ("Disposizioni inerenti l'attività di protezione civile nell'ambito dei bacini in cui siano presenti dighe" Circular Letter n° DSTN/2/7019).

- Hydraulic Assessment (*"Verifiche Idrauliche" Circular n. 3199 of the Dam Autority*) – second to the Circular the dam owner is obliged to execute updated hydrological analyses for the evaluation of the maximum floods corresponding to increasing return period and the corresponding assessment of the hydraulic safety of the dams.

- Environmental code D.lgs. 152/2006 art. 114 Dams - Dighe. A siltation managing plan is requested for each reservoir to assure maintenance of the dam and appurtenances according to the water quality of the downstream river. The Regional department approves the plan on the advice of the National Dam Authority

State/ Country	Law/Act concerning dam safety	Decrees etc. concerning dam safety
Italy	Yes	Yes
NOTE: Tailings dams are in domain of the "Ministry of Industry Ministry" (Mining Section) and no specific Regulation is available for them.		

19.1.2. Responsibilities for dam safety

The Owners are responsible for the dam safety. About 60% of the Italian dams are owned by private companies, the remaining 40% of dams are owned by public entities (public "Consortiums" or Companies, Municipalities, State). To assure greater safety and the proper operation of the dam, a *"Responsible Engineer"* must be appointed by the owner, for each large dam in operation.

The owner is obliged to carry out monitoring of the dam. The monitoring system is approved by the administration (all dams are provided with more or less complete monitoring systems).

Each six months, the owner must send to the Dam Authority an "*Attestation*" or a certificate by the Responsible Engineer for each dam which includes the results and diagrams of the main measurements, that confirms the safety of the dam (good conditions of the dam and its operation).

19.1.3. Arrangements for independent dam safety supervision

The supervision on safety of dams depends on size of the dam. The dams are divided in two basic groups (large dams and small dams) – the safety of dams in each group is organized by different "authority":

- Large dams - Supervision of all the phases from design to operation of a large dam as well as participation in the updating of Regulation and technical standards is performed by the National Dams Authority. The NDA carries out the supervision of the surveillance and control activities carried out by the owner during the operation twice a year

- Small dams - Approval of the concessions as well as projects and supervision of the activities relevant to the construction and operation of "small dams" is performed by Regional Administrations.

The control and surveillance activities (inspections, monitoring,) to be carried out for the structures (dam, foundation, reservoir slopes, appurtenant works, etc.) are specified in a document named "*Conditions for Operation and Maintenance*". The sheet defines the type, extension, frequency, etc. of each surveillance activity and it is issued by the Dam Authority for each singular dam (the document must be subscribed also by the dam owner as an obligation document). Every month, the owner must send to the Dam Authority a list of the results of the monitoring and observations. The observations must be also listed in a register at the dam site.

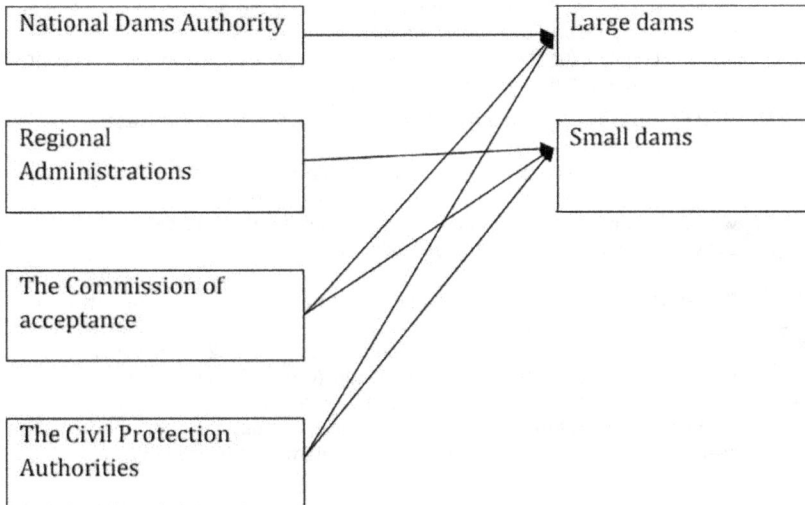

19.2. DAM CLASSIFICATION

19.2.1. General principles of classification

The classification criteria are based mainly on characteristic of dam (height, spillway capacity, proximity of other dams, etc.). These criteria do not apply to tailings dam. Tailings dams are in domain of the "Ministry of Industry Ministry" (Mining Section) and no specific Regulation is available for them.

A classification is done according to the dimensions of the dams. There is no classification according to risk or hazard criteria.

Number of classes: 2

19.2.2. Dam classes overview

Dams subjected to the National Dam Authority

- Dams with height H>15m, or

- Dams with reservoir volume V> 1.000.000 m^3

Dams assigned to the Regional Authorities (21 regions)

- Dam that do not fulfil the above listed criteria

19.3. TECHNICAL FRAMEWORK

- "Norme Tecniche per la progettazione e la costruzione delle dighe" (Technical Rules for the Design and Construction of Dams), D.M. LL.PP. n°44. - This law defines the technical rules for the design and construction of the dams and it considers only new dams (no directions are given about existing dams already in operation). Specific rules are given for the different dam types: gravity, hollow gravity, buttress, arch, multiple arches, gate-structure, earth and rockfill dams.

  NOTE: A complete update of this Regulation is currently to release.

- Sistemi di allarme e segnalazioni di pericolo per le dighe di ritenuta" (Warning and Alarm Systems for Dams), Ministry of Public Works, Circular n° 1125 - installation of warning signs along the stream, downstream the dam, to alert. All the above stated activities must be carried out by the dam owner.

- "Prescrizioni inerenti l'applicazione del regolamento dighe approvato con DPR n° 1363 del 1959" ("Directions concerning the application of the Regulation about Dams n° 1363/1959"), Ministry of Public Works, Circular n° 352. - The Circular updates some specific elements of the 1959 Regulation (doc. n.2), among which the dam break analysis to determine the flood propagation and the inundation maps, the "Foglio di Condizioni per l'Esercizio e la Manutenzione" ("Sheet of Conditions for Operation and Maintenance"), a classification of the alert/alarm conditions is given.

- Indirizzi operativi per la gestione organizzativa e funzionale del sistema di allertamento nazionale e regionale per il rischio idrogeologico ed idraulico ai fini della protezione civile" (Operational directions for the management of the national and regional alert systems for the hydro-geological and hydraulic risk, for Civil Protection purposes), Presidency of the Council of Ministers.

There are no legal guidelines.

19.4. SUPPLEMENTARY REFERENCES

More information about dam safety management, legislation and classification can be found in [1] and [7].

[1] http://www.itcold.it/index.asp

[2] http://en.wikipedia.org/wiki/Italy

[3] http://unibg.academia.edu/MarcoLazzari/Papers/673864/The_role_of_AI_technology_in_the_management_of_dam_safety_the_DAMSAFE_system

[4] Jansen, R.B., Dams from the beginning (ussdams.com/ussdeducation/.../damsfrombegin.doc)

[5] Mazza,G., Giuseppetti, G., Ruggeri, G., Bonaldi, P. – Integrated monitoring systems for the seismic reassessment of existing dams - Dam Safety: Proceedings of the International Symposium on New Trends and Guidelines on Dam Safety, Barcelona, Spain, 17–19 June 1998 p.p.1051–1057

[6] http://palnet.cmscalve.bg.it/gleno%20-%20convegno/atti%20convegno/Ing.%20Giovanni%20Ruggeri%20-%203%B0%20intervento/Files/gleno-Ruggeri.pdf

[7] http://www.ors.regione.lombardia.it/resources/pagina/N1201fae8fed5b5be435/N1201fae-8fed5b5be435/C_2_ContenutoInformativo_1111_ListaAllegati_Allegato_8_All_Allegato.pdf

20.   JAPAN

Reporter: Hiromi Kotsubo

20.1.   MAIN PRINCIPLES OF DAM SAFETY MANAGEMENT

Japan is a unitary parliamentary democracy and a constitutional monarchy where the emperor has a limited power mainly restricted in greater part to ceremonial duties. Japan consists of forty-seven prefectures, each overseen by an elected governor, legislature and administrative bureaucracy. Each prefecture is further divided into cities, towns and villages.

The total number of large dams in operation in Japan is 3076 (Register of dams 2012), and 2732 of these are dams higher than 15 meters. The number of dams is increasing – currently more than 40 dams are being constructed. Besides for the integrated river basin development, the dams/ storage facilities are developed mainly for irrigation, flood control and water supply.

20.1.1. Legal framework for dam safety

There are several national acts and ordinances covering aspects of dam safety, but the paramount requirements are given in the River Law no. 2005.7.29 issued by the Government.

To enforce the River law several ordinances have been issued by the Government (Ordinance for Enforcement of the River Law 2005.6.1, Ordinance for Structural Standard for River Administration Facilities) and the Ministry of Land, Infrastructure, Transport and Tourism (Ordinance for Enforcement of the River Law 2008.9.1). The extent and frequency of inspections is specified in dam inspection regulation (1998.1.23) issued by the Ministry of Construction instructions.

There are additional regulations about water-power generation issued by Ministry of Economy, Trade and Industry and a standard about land improvement issued by Ministry of Agriculture, Forestry, and Fisheries.

The dams that are not located in a river zone are treated on the standard and regulation basis as the dams situated in the river zone.

State/ Country	Law/Act concerning dam safety	Decrees etc. concerning dam safety
Japan	Yes	Yes

20.1.2. Responsibilities for dam safety

The dam owners are responsible for the safety of a dam (of the structure itself) from the design phase to operation phase (proper operation) and decommissioning of a dam. The owner is also responsible to organize and carry out regular inspections and dam surveillance programs. Second to the River Law it is necessary that a facility's management, regulation and operation are specified in detailed rules which must be defined for each single dam. The responsibility for these operational rules is in the domain of River administrators. The River administrators provide the operational rules – instructions and control the adequacy of them.

The dam owners can be private parties (persons/companies), municipalities, counties or the State itself.

20.1.3. Arrangements for independent dam safety supervision

The safety supervision of dams in Japan is organized in different levels. The first level of supervision of the dam owner's work is performed by the dam inspector. The second level of supervision is performed by the river office and the district office. The ultimate dam safety authority supervision is carried out by The Ministry of Land, Infrastructure, Transport and Tourism (representing the government).

These multiple levels of surveillance and review (dam guardian, experienced engineer, expert(s), federal authorities) allow minimizing the risk of having an undetected critical feature or a potential hazard.

20.2. DAM CLASSIFICATION

20.2.1. General principles of classification

The classification criteria are based mainly on purpose and type of dam and are issued in Government Ordinance for Enforcement of the River Law. There is no classification based on risk and loss of life or damage caused on property. Criteria for classification are the characteristic of dam (height, spillway capacity, proximity of other dams, etc.). The same criteria apply to all types of dams and reservoirs, also to those which are not located in a river zone.

Dams are classified by Government Ordinance for Enforcement of the River Law and by the notification from Director General of the River Bureau.

Number of classes: 2

20.2.2. Dam classes overview

The dams in Japan are categorized by two categorisations.

The classification by the Government Ordinance for Enforcement of the River Law divides the dams in two categories:

- A dam that has spillway gates and has an impoundment sections whose total length is 10 km or more.

- In case there are two or more dams located not more than 30 km apart along the same river and the total length of impoundment sections associated with those two or more dams is 15 km or more, one of the dams has spillway gates.

- A dam other than those listed in the two preceding items whose height from foundation ground to overflow crest is 15m or more.

The classification by the notification from Director General of the River Bureau also divides the dams in two categories, but is using slightly different criteria:

- A dam which downstream flood discharge increases more remarkably than before, and the dam needs to cope with a flood by storing increased flood discharge in a dam lake (mainly large-scale dam only for power generation)

- A dam which needs to cope with a flood by lowering the water level beforehand for flood damage prevention by upstream water level rise at the time of a flood (mainly built in the middle part of large river)

- A dam which is appropriate to lower a water level, as the discharge capability of spillway gates are larger than reservoir capacity or the operation method of a spillway gates are complicated. (mainly dam with many gates)

- A dam which does not have the influence on the downstream area by flood discharge, even if the water level of the dam is at normal water level (mainly a small-scale dam, a multi-purpose dam, and a flood control dam)

The dams are also categorized unofficially by ownership and by purpose:

- governmental; Ministry of Land, Infrastructure, Transport and Tourism, Ministry of Agriculture, Forestry, and Fisheries,

- public Japan Water Agency,

- owned by a local government: prefectures, cities, and so on,

- owned by electric power companies,

- owned by private enterprises,

- purpose; multi-purpose and single-purposed dams for flood control, water utilization water supply, irrigation water, power generation etc.

There is no categorization regarding the risk.

20.3. TECHNICAL FRAMEWORK OVERVIEW

Technical framework that concerns dam safety:

- Government Ordinance for Structural Standard for River Administration Facilities 2000.6.7 last revision (1976.7.20)

- Ministry of Construction instructions – for a dam inspection 1998.1.23 last revision (1968.2.17)

20.4. SUPPLEMENTARY INFORMATION

More information about dam safety management, legislation and classification can be found in [1] and [2].

[1] Yamaguchi, Y., Kobori, T., Sakamoto, T., (2010). Safety management and seismic safety evaluation for dams in Japan Paper presented at the ICOLD European Club Symposium, Innsbruck, Austria 2010.

[2] http://www.mlit.go.jp/tochimizushigen/mizsei/water_resources/contents/policy.html

21. MEXICO

Reporters: Felipe I. Arreguín-Cortes / Rodrigo Murillo-Fernández

21.1. MAIN PRINCIPLES OF DAM SAFETY MANAGEMENT

Mexico is a federation with 31 states and one federal district (Mexico City).

There are 4900 dams in Mexico of which 813 are large dams according to the ICOLD criteria. The main purposes of large dams are irrigation and generation [5].

21.1.1. Legal Framework for Dam Safety

Rivers and their waters are mainly national owned. Rivers, riverbeds and flood plains are regulated by the Law on National Waters [4], which establishes regulations for the use of water and land. The Law on National Waters is currently under review for amendment in the House of Representatives (2012).

Regulations in Mexico concerning dam safety and related items are:

- Article 27 of the Constitution of Mexico, 1917

- Law on National Waters, 2004 and By-Laws, 2002 [4]

- Civil Protection Act, 2000

- Agreement by which those responsible for dams in operation are identified, 2010 [6]

The following table gives an overview of dam safety legislation in Mexico:

State/ Territory	Law/Act concerning dam safety	Decrees etc. concerning dam safety
Mexico	Yes	Yes
States	No	No

21.1.2. Responsibilities for Dam Safety

The owners or operators of dams are liable for damage caused by the break or misuse of these works.

In case of imminent danger to people, Federal, State or Municipal Civil Protection Systems may intervene to reduce the risks.

21.1.3. Arrangements for Independent Dam Safety Supervision

National Water Commission or CONAGUA (a federal agency) is responsible for surveillance of condition and behaviour of national rivers and dams of its property. There is a particular office in CONAGUA charged of that, but this agency does not inspect the dams of other owners or users.

21.2. DAM CLASSIFICATION

Dams are all retaining structures with reservoirs over 250.000 m³. The categories mainly used, but not standardized, to distinguish dam size are:

- Large Dams: h > 15 m

- Medium Dams 10 < h < 15 m

- Small Dams 5 < h < 10 m

- Dikes h < 5 m

Dams are categorized into normally water-filled reservoirs, large and small dry flood protection reservoirs, weirs and tailings dams.

For risk level description the USBR (1988) criteria are used:

- High. The failure is very likely to happen and in case of failure, the damages include loss of human life or serious environment damages.

- Medium. The failure of the work is likely to happen and in case of failure, the damages would be mainly material and limited to the environment.

- Low. The failure of the work is unlikely to happen and in case of failure, the damages would be small and limited to the work.

- Null. No risk of failure.

21.3. TECHNICAL FRAMEWORK OVERVIEW

CONAGUA follows an internal Dam Safety Program, which is based on USBR Program. Also the Federal Electricity Commission has a particular procedure to assess the safety of dams of its property and under its responsibility.

- Manual para la capacitación en Seguridad de Presas, Comisión Nacional del Agua (Conagua), 2000 and 2001, Spanish translation of "Training aids for Dam Safety" originally published by U. S. Bureau of Reclamation, 1988.

- Procedure for the preparation and submission of reports of dam safety, 2011, for Conagua staff.

- Instructions for obtaining basic information from dams, 2011, for Conagua staff.

Currently, dam owners are not obliged to have a dam safety surveillance program. Current practice consists of visual inspection of the condition of the structures in case of serious physical or hydraulic malfunction, further studies and deeper analyses if needed.

21.4. REFERENCES

[1] Arreguín F, Herrera C, Marengo H, Paz Soldán G, "El desarrollo de las presas en México (The development of dams in Mexico)", editorial: Asociación Mexicana de Hidráulica, Volume 5, First edition, 1000 copies, 224 pages, ISBN: 968-7417-28-5, Mexico, 1999.

[2] Paz Soldán G, Marengo H. y Arreguín F, "Las presas y el hombre (Dams and the man)", Asociación Mexicana de Hidráulica e Instituto Mexicano de Tecnología del Agua, ISBN-968-5536-49-X, Mexico, 2005 May.

[3] Arreguin F. "Cavitación y aireación en obras de excedencia (Cavitation and aeration in spillways)", Asociación Mexicana de Hidráulica e Instituto Mexicano de Tecnología del Agua, Mexico, 2005 May.

[4] Diario Oficial de la Federación "Ley de Aguas Nacionales y su Reglamento (Law on National Waters and By-Laws)" ISBN 968-817-626-5, Conagua, Mexico, 2004.

[5] Informatics System of Dam Safety (National Inventory of dams), Conagua, Mexico, 2012.

[6] Diario Oficial de la Federación "Acuerdo mediante el cual se identifica a los responsables de las presas en operación (Agreement by which those responsible for dams in operation are identified)", Mexico, 27 May 2010.

22.   NETHERLANDS

Reporter: Hans Janssen

22.1.   MAIN PRINCIPLES OF DAM SAFETY MANAGEMENT

The Netherlands, a constituent country of the Kingdom of the Netherlands, is divided into twelve administrative regions (provinces). All provinces are divided into 408 municipalities. The monarch is the head of state, but that position is equipped with limited powers. The executive power is formed by the Ministerraad, the deliberative council of the Dutch cabinet. The cabinet is responsible to the bicameral parliament, the States-General, which also has legislative powers.

The country is also subdivided in 26 water districts, governed by a water board (waterschap or hoogheemraadschap), each having authority in matters concerning water management. The 26 water districts cover also about 100 areas; all protected by a system of primary water defences, a so-called "dike ring".

The Netherlands is a low-lying country along the North Sea, in the delta of the rivers Rhine, Meuse and Scheldt. Protection against flooding started in the Middle Ages with small dikes to protect low-lying areas and to reclaim land from flooding. Nowadays the country is protected against threats coming from the sea and the rivers by an extensive system of primary water defences. The man-made water defences consist of a system of dikes, dams and barriers along the major rivers, the estuaries en large lakes like Lake IJssel and Lake Marken. Besides this, the strength of the natural dune defences along the coastline is actively maintained by men. The total length of the primary water-defence system is over 3500 km.

According to the ICOLD Register of dams 12 of the existing dams and barriers are considered as large dams.

22.1.1. Legal framework for dam safety

Legal safety standards for primary water defences including dams were first defined and included in the Flood Defences Act (in Dutch: "Wet op de Waterkering") of 1996.

In 2010, the Flood Defences Act was integrated into the Water Act ("Waterwet").

Safety standards as included in the Water Act [4] range from a maximum allowed overtopping probability of 1/10.000 per year for densely populated dike-ring areas along the coast, to 1/250 along the upstream Dutch part of the Meuse river. Most dams are not part of a dike ring, but rather connect different dike rings as they serve to close (former) estuaries. The safety standard of dams is generally equal to the highest safety standard of the dike rings behind it.

The Water Act not only specifies safety standards, but also prescribes a regular safety assessment for all primary water defences.

State/ Country	Law/Act concerning dam safety	Decrees etc. concerning dam safety
The Netherlands	Yes	Yes

A number of policy changes are underway, but not yet fully included in legislation:

- Safety standards will most probably be revised, and based on an integral flooding probability (and actually on flood risk by making the safety standard also dependent on the protected value at risk) rather than just an overtopping probability

- The Water Act is likely to be integrated into an overall Environmental Act ("Omgevingswet")

- Government, water boards and provinces have agreed to decrease the frequency of dike assessments (once per 12 rather than once per 6 years) and to limit the degree to which dike reinforcements are subsidized (from 100% to 50%). Also, the monitoring of compliance with the Water Act will be simplified from a two-stage-process (water defence responsible, province, state) to a single-stage direct monitoring by the Human Environment and Transport Inspectorate (ILT) of the Netherlands.

22.1.2. Responsibilities for dam safety

The responsibility for dam safety is centralised and lies with the Ministry of Infrastructure and Environment. The responsibility is divided into three bodies:

- The Directorate-General for Spatial Development and Water Affairs (DGRW) [5] is responsible for preparation of the legislation, safety standards, guidelines, safety assessments and hydraulic boundary conditions for safety assessments. The directorate is also responsible for the budget to reinforce water defences that do not comply with the safety standards.

- The Directorate General Rijkswaterstaat is responsible for implementation of guidelines and boundary conditions for safety assessments into practice. The Rijkswaterstaat is also authorized by the Ministry for some of the operational aspects of the dams - Management and maintenance of the coastline, rivers, main waterways and 10% of the defences (mainly dams and barriers). The remaining 90% of the defences are managed and maintained by Water boards. They are responsible for management and maintenance of water barriers (dunes, dikes, quays and levees), management and maintenance of local waterways, maintenance of a proper water level in polders and local waterways and maintenance of surface water quality through wastewater treatment.

- The Human Environment and Transport Inspectorate (ILT) is responsible for inspection and evaluation of safety assessment and management.

The Ministry is responsible for legislation, but "concedes" the operational aspect of the dams to the Governmental Agency (Directorate General Rijkswaterstaat) and the Water Boards. The Agency and the Water Boards are responsible for operation, maintenance and management of the dams. For every part of the water defence system they are obliged (by law) to establish a map with the exact location of the dam, a ledger, describing the characteristic of the water defence, and a technical register of maintenance works (refurbishment, reinforcement of weak areas) needed to maintain the required safety. Every 6 years (to become every 12 years) a safety assessment should be performed, which is checked by theILTT Inspectorate.

The Ministry is also responsible for providing a Legal (Dike Safety) Assessment Instrument and Guidelines, but in practice these are published by the Helpdesk Water (www.helpdeskwater.nl) and the Expertise Network on Flood Defences ENW (www.enwinfo.nl) respectively. The ENW also has an important role in the quality assurance of technical documents related to the Legal Dike Safety Assessment.

22.1.3. Arrangements for independent dam safety supervision

The periodic safety assessments are controlled and reviewed by the Ministry of Infrastructure and the Environment. If the achieved safety is insufficient the Ministry orders the agency to draw up the plans and to carry out the maintenance works and the necessary refurbishment to increase the safety to the required level.

Each dike ring area has a safety standard that depends on the nature of the possible flooding and the scale of the potential damage that a failure could cause in the dike ring. These standards are expressed as an exceedance frequency and determine the conditions that the water defence should withstand. For dams the safety level is adapted to the highest safety level of the dike rings it is connecting.

22.2. DAM CLASSIFICATION

22.2.1. General principles of classification

There is no dam safety categorisation. All dams, dikes and barriers are treated the same way as the primary water defences, which are all shown on the maps in Annex I of the Water Act.

22.3. TECHNICAL FRAMEWORK OVERVIEW

The technical framework consists of design guidelines, standards and technical reports related to water defence safety and also of a safety assessment regulation and the hydraulic boundary conditions for the assessment. Both the safety assessment regulation and the hydraulic boundary conditions for assessment are updated every 6 years (to become every 12 years) for the next periodic safety assessment.

The technical framework consists of the Legal Dike Safety Assessment Instruments, Technical Reports and Guidelines. The Guidelines also consider design and maintenance aspects. Due to the upcoming modification (to a new type) of safety standards, many of these documents will undergo substantial changes in the years to come.

The Legal Dike Safety Assessment Instruments are documented in two key documents:

- The so-called HR-Book or Hydraulic Boundary Conditions Book

- The so-called VTV or Flood Defence Assessment Rules

At present, the version of 2006 is still prescribed, which can be found at http://www.helpdeskwater.nl/onderwerpen/waterveiligheid/primaire/toetsen/wti2006-vigerend/ (Dutch versions only). One can also search for:

- Hydraulische Randvoorwaarden voor primaire waterkeringen, Ministerie van Verkeer en Waterstaat, ISBN978-90-369-5761-8, Augustus 2007 (National Hydraulic Boundary Conditions for Water Defences) (the so-called HR-Book)

- Voorschrift Toetsen op Veiligheid Primaire Waterkeringen Hydraulische Randvoorwaarden voor primaire waterkeringen, Ministerie van Verkeer en Waterstaat, ISBN978-90-369-5762-5, September 2007 (Flood Defence Assessment Rules) (the so-called VTV)

In some cases, simple assessment rules are not enough to establish that a flood defence meets the safety standard. In that case, more complex assessments can be made using so-called Technical Reports. These reports are provided by the Expertise Network on Flood Defences (ENW). The ENW also provides Guidelines ("Leidraden") which can be used for design and maintenance purposes, and a framework ("Fundamentals" or "Grondslagen") document describing the overall philosophy behind all ENW Guidelines and Technical Reports. The original versions of these ENW documents are given on www.enwinfo.nl; some English translations of key documents are given on http://www.enwinfo.nl/asp/uk.asp?DocumentID=112&niveau=1

Some key documents from ENW are given below. They can be downloaded from www. enwinfo.nl, when English versions are available, direct weblinks to their translations are given as well:

Fundamentals:

- Grondslagen voor Waterkeren, Ministerie van Verkeer en Waterstaat, TAW, ISBN: 90-369-373-5-3, A.A. Balkema Uitgevers B.V. (also available in English: Fundamentals on Water Defences, www.enwinfo.nl/engels/downloads/FundamentalsWaterDefences.pdf)

Guidelines:

- Leidraad Rivieren, Ministerie van Verkeer en Waterstaat, ENW, Den Haag, juli 2007 (Guide on River Dikes)

- Leidraad Zee- en Meerdijken, Ministerie van Verkeer en Waterstaat, TAW, December 1999 (Text also available in English: Guide on Sea and (large) Lake Dikes; www. enwinfo.nl/engels/downloads/GuideSeaandLakeDikes.pdf)

- Leidraad Waterkerende Kunstwerken en Bijzondere Constructies, (in Dutch), TAW, juni 1997 (Guide on Water Retaining Hydraulic Structures and Special Objects)

Technical Reports:

- Technisch Rapport Waterkerende Grondconstructies, TAW, P-DWW-2001-035, ISBN 90-369-3776-0, juni 2001 (Text also available in English: Technical Report on Water retaining Earth Structures; www.enwinfo.nl/engels/downloads/TRSoilStructures.pdf)

22.4. SUPPLEMENTARY INFORMATION

More information about dam safety management, legislation and classification can be found in:

[1] http://www.ifi-home.info/isfd4/docs/May8/Session_1030am/Room_E_1030/Safety_Assessment_of_primary_flood_defences_in_the_Netherlands.pdf

[2] Berga, L. – Dam Safety: Proceedings of the International Symposium on New Trends and Guidelines on Dam Safety, Barcelona, Spain, 17–19 June 1998 p.p.345–352

[3] http://en.wikipedia.org/wiki/Flood_control_in_the_Netherlands

[4] English version of helpdesk Water website for information on water management and water legislation: http://www.helpdeskwater.nl/algemene-onderdelen/serviceblok/english/

[5] http://www.government.nl/ministries/ienm/organisation

[6] Flood risk and water management in the netherlands, a 2012 update, rijkswaterstaat report WD0712RE205, aug2012, http://www.helpdeskwater.nl/algemene-onderdelen/serviceblok/english/water-and-safety/@34443/flood-risk-and-water/

23.   NEW ZEALAND

Reporter: Peter Mulvihill

23.1.   MAIN PRINCIPLES OF DAM SAFETY MANAGEMENT

New Zealand is a parliamentary democracy with a central government and central legislation. The role of regulator is the part of the function of a government department called the Department of Building and Housing. Many of the regulatory responsibilities for dam safety are assigned to regional councils which are river catchment/ regionally based and in some cases to unitary authorities which have a combined territorial and regional role.

There are estimated to be 1,150 dams in New Zealand that fall under the dam safety regime.

23.1.1. Legal framework for dam safety

There are two national acts covering aspects of dam safety; the Resource Management Act (1991) and Building Act (2004), as well as Building Regulations (Dam Safety) (2010). The Civil Defence Act (1989) may apply for emergency situations. In addition, there is a separate act for design of tailings dams – Hazardous Substances and New Organisms Act (1996).

The dam safety legislation applies to all dam owners, independent of ownership (public and private owners).

State/ Country	Law/Act concerning dam safety	Decrees etc. concerning dam safety
New Zealand	Yes	Yes

23.1.2. Responsibilities for dam safety

The dam owners are responsible for the dam safety.

23.1.3. Arrangements for independent dam safety supervision

Governmental safety supervision of dams is conducted by the regional councils or unitary authorities.

23.2.   DAM CLASSIFICATION

23.2.1. General principles of classification

The dam classification system is based on consequences, which is determined through the damage level and public safety.

Damage level is based on:

- No. of residential houses destroyed/damaged

- Extent of damage and time to restore to operation critical or major infrastructure

- Damage to natural environment

- Community recovery time

Public safety is measured through population at risk.

Number of classes: 4

23.2.2. Dam classes overview

Classification system is based on the potential impact of the dam failure. These are based on incremental loss of life AND socioeconomic, financial and environmental impacts – High, Medium, Low, Very Low. (Note: While the NZSOLD Dam Safety Guidelines include a Very Low Classification the legislation only includes High, Medium and Low).

23.3. TECHNICAL FRAMEWORK OVERVIEW

All dams are covered by the Building Act 2004 and need to meet the Building Code standards. Only large dams (>20.000 m^3, >3m depth) are covered by the regulations.

Guidelines, norms and standards on dam safety are developed by New Zealand Society on Large Dams (NZSOLD).

NZ Dam Safety Guidelines (2000) are currently being revised to be consistent with the new regulations (Building Regulations, Dam Safety, 2010).

23.4. SUPPLEMENTARY INFORMATION

More information about dam safety management, legislation and classification can be found in:

[1] New Zealand Building Act 2004
[2] New Zealand Building (Dam Safety) Regulations 2008

24.   NIGERIA

Reporter: Imo Ekpo

24.1.   MAIN PRINCIPLES OF DAM SAFETY MANAGEMENT

Nigeria is a federal constitutional republic comprising 36 states and its federal capital territory. There are about 200 dams of varying size that have been constructed for multiple purposes.

24.1.1. Legal framework for dam safety

The issue of dam safety is regarded very seriously. The following regulatory documents exist for that purpose:

- Water Resources Act 101 of 1993.

- Rules and Regulations for the Administration and enforcement of the Water Resources Act No 101 of 1993.

- By Law for Regulation, Monitoring and Supervision of Dams and Reservoirs in Nigeria.

State/ Federal State/ Province/Territory	Law/Act concerning dam safety	Decrees etc. concerning dam safety
Nigeria	yes	yes
States	no	no

24.1.2. Responsibilities for Dam Safety

Dam owners are directly responsible for the safety of their dams.

There is no state/governmental supervision of dam safety in Nigeria. A Dam Safety Division exists in the Federal Ministry of Water Resources. This Division liaises with the dam owners to assess the dam status and make recommendations or occasionally intervene.

In practice unfortunately the Water Law and Dam Safety regulatory issues are not effectively operational. The legal office in Nigeria (Federal Ministry of Justice) is currently harmonizing the Water Law.

24.2. DAM CLASSIFICATION

The categorization is partly based on ICOLD standards:

- Large Dams: $H \geq 15$ m

- Medium Dams: 8 m $< H < 15$ m

- Small Dams: $H \leq 8$ m and $V < 5$ million m^3.

24.3. TECHNICAL FRAMEWORK OVERVIEW

There is no technical framework concerning dams and dam safety in Nigeria.

25.   NORWAY

Reporter: Grethe Holm Midttømme

25.1.   MAIN PRINCIPLES OF DAM SAFETY MANAGEMENT

Norway is a unitary state with 19 administrative counties (fylker). The King and the government are represented at the county level by a County Governor.

There are more than 3,100 registered dams in Norway, and 335 of these are dams higher than 15 meters. There are probably also several thousands of small unregistered dams. The number of registered dams is increasing due to ongoing registration of existing dams and planning and construction of new small hydro power projects.

25.1.1. Legal framework for dam safety

There are several national Norwegian acts covering aspects of dam safety, but the paramount requirements are given in the Act no 82 of 24 November 2000 (Act of water resources and ground water) issued by the Ministry of Petroleum and Energy.

Since 2010 specific requirements about dam safety have been assembled in "Regulations governing the safety of watercourse structures" (reg. no. 1600 given by royal decree on 18 December 2009), also denoted as the Dam Safety Regulations. The requirements in the Dam Safety Regulations are consequence based, which means that the most comprehensive safety requirements apply only to the dams in the highest consequence classes. Only a few requirements apply to dams in consequence class 0.

State/ Country	Law/Act concerning dam safety	Decrees etc. concerning dam safety
Norway	yes	yes

25.1.2. Responsibilities for dam safety

The dam owners are responsible for the dam safety (from design phase to operation phase/ until decommissioning). Dam owners can be private (persons/companies), municipalities, counties or State.

The dam safety authority (regulator) is The Norwegian Water Resources and Energy Directorate (NVE).

25.1.3. Arrangements for independent dam safety supervision

Governmental safety supervision of dams in Norway has been carried out by NVE since 1909. Until 1981 governmental supervision was restricted to dams with concession in the law, but thereafter governmental supervision was extended to all dams which could pose a hazard to people, property or the environment.

Since the 1990's dam safety management in Norway has been based on the internal control principle which in practice means that the dam owner has to document that all his dam safety activities are done in accordance with the Water Resources Act and the Dam Safety Regulations. Requirements about internal control for dam safety management are given in a separate regulation issued in 2003 and revised in 2011. NVE controls the fulfillment of these requirements through frequent audits of the dam owner's organization and documents, including site visits. NVE also controls and gives approval to plans for construction of new dams and reassessment and plans for upgrading of existing dams.

The Dam Safety Section in NVE has 20 employees, mostly dam safety specialists, and each of them is responsible for safety supervision of dams and dam owners in a geographical region.

25.2. DAM CLASSIFICATION

25.2.1. General principles of classification

Criteria for classification are life safety (loss of life or population at risk), infrastructure, property (direct economic loss to third parties) and environmental impacts. Loss of infrastructure, such as water supply, power supply and roads, are assessed with respect to the importance of the actual infrastructure to society. Cultural heritage loss is considered as part of environmental impacts. Any secondary effect caused by long duration and/or intensity of flood wave must also be considered (typically additional risk of erosion and landslides).

The same criteria apply to all types of dams and reservoirs.

Dam owners are responsible for classification when planning a new dam or reconstruction of an existing dam, and for existing dams; if the dam is not classified before, or if there are changes that may affect the class (for example development of a new building area downstream of the dam). The classification must be approved by NVE (the dam safety authority) unless the dam is very small (see below about limit values for class 0).

Number of classes: 5

25.2.2. Dam classes overview

All dams are classified in one of the following consequence classes:

- Class 4: extra high consequences, i.e. more than 150 housing units affected, possibly also in combination with substantial damage to infrastructure, property and the environment.

- Class 3: high consequences, i.e. 21–150 permanent housing units affected, or damage to essential infrastructure (main road/railway with heavy traffic, very important water supply or power supply installation etc.), or substantial damage to property and environment.

- Class 2: medium consequences, i.e. 1–20 permanent housing units affected, or damage to infrastructure of importance, or major damage to property and environment.

- Class 1: low consequences, i.e. no permanent housing units affected but possible damage to cabins, camping sites, sports grounds etc. (areas where people may stay for shorter time periods, equivalent to 0–1 housing unit), or damage to infrastructure of local importance, or damage to property or environment.

- Class 0: insignificant consequences, and all dams less than 2 m high and less than 10,000 m³ reservoir volume (both criteria fulfilled).

The criteria for class 1–4 are given in the table:

Class	Consequences		
	Housing units affected	Damage to infrastructure	Damage to property and environment
4	>150		
3	21–150	Substantial	Substantial
2	1–20	Medium	Medium
1	<1	Low	Low

If one of the consequence criteria in the table is fulfilled, the dam class must be set to the class given in the table corresponding to the single criterion.

If several criteria are fulfilled for a specific class, and the sum of consequences is extra high, NVE can put the dam in a higher consequence class than given directly by the separate criteria.

Dams with height < 2 m and reservoir volume < 10,000 m³ can be classified in class 0 without notification to and approval from NVE. Larger dams may also be placed in class 0 if failure consequences are insignificant.

25.3.   TECHNICAL FRAMEWORK OVERVIEW

Details and practical solutions to the requirements given in the Dam Safety Regulations are given in central technical guidelines (edited and issued by NVE, but with contributions from research institutions, consultants, dam owners and others). By March 2013, there are 11 guidelines:

- Guideline for flood calculations (NVE, 2011)

- Guideline for planning and construction (NVE, 2012)

- Guideline for inspection and reassessment (NVE, 2002)

- Guideline for concrete dams (NVE, 2005)

- Guideline for spillways (NVE, 2005)

- Guideline for masonry dams (NVE, 2011)

- Guideline for surveillance and instrumentation (NVE, 2005)

- Guideline for determination of loads (NVE, 2003)

- Guideline for dam break flood analysis (NVE, 2009)

- Guideline for embankment dams (NVE, 2012)

- Guideline for penstocks, gates and valves (NVE, 2011)

In addition, there are some other relevant guidelines concerning the safety of small dams/ class 0 – dams, for example:

- Guidelines for planning, construction and operation of small dams (NVE, 2006)

25.4. SUPPLEMENTARY INFORMATION

More information about dam safety management, legislation and classification can be found in [1] and [2].

[1] Midttømme, G. Holm. Grøttå, L. and Hyllestad, E. (2010). New Norwegian Dam Safety Regulations. Paper presented at the ICOLD European Club Symposium, Innsbruck, Austria 2010.

[2] www.nve.no > information about dam safety management, legislation and classification in Norway (in Norwegian).

26.   PERU

Reporter: Miguel Suazo

26.1.   MAIN PRINCIPLES OF DAM SAFETY MANAGEMENT

Peru, officially the Republic of Peru, is a representative democratic republic divided into 25 regions. Federal legislation rules dam safety.

26.1.1. Legal framework for dam safety

Supreme Decree 253–72 of 19th March 1972 was issued by the Ministry of Agriculture / Directory of Waters and Irrigation / Direction of Watering Infrastructure. It includes standards for inspection, operation and maintenance of dams and reservoirs.

State/ Country	Law/Act concerning dam safety	Decrees etc. concerning dam safety
Peru	yes	yes

26.1.2. Responsibilities for dam safety

Dam safety control is the responsibility of the owners. The owner is obliged to have a surveillance program.

26.1.3 Arrangements for independent dam safety supervision

Dam safety supervision is made by governmental organisms. There is not a unique authority; instead each sector has a responsible area. The dams belonging to private sector report depending on the sector they serve.

For dams used for energy and mining, OSINERGMIN (organism of investment in energy and mining) is the responsible body for the supervision. For special projects of the agriculture sector, the supervision is made by INADE (national institute of development). The control entities make periodical inspections.

26.2.   DAM CLASSIFICATION

Dams are classified according to the criteria given by ICOLD.

They are classified mainly as water supply dams (agriculture, energy, industry, water supply) and tailing dams.

There is no official classification in relations to risks, even though risks are mentioned for the definition of inspection priorities.

26.3. TECHNICAL FRAMEWORK OVERVIEW

Standards for inspection, operation and maintenance of dams and reservoirs are given in the Supreme Decree 253–72 mentioned above. The standards have the following content:

I - INSPECTION OF DAMS AND RESERVOIRS

Purpose

Scope

Responsibility for the inspection

Type of inspections

Suitable periods for inspections

Proceedings for inspections

Corrective actions

II - OPERATIONS OF DAMS AND RESERVOIRS

Distribution of water

Operation and measurement equipment

Rules for the operation

III - MAINTENANCE OF DAMS AND RESERVOIRS

Justification

Responsibilities

Period after transference

Responsibility of executive offices and other entities

Responsibility of specialized offices of the general directory

Responsibility for maintenance

Control of founding's

Need to follow the given instructions

Seepage

Cracks

Erosion

Slope stability

Inspection of equipment

Earthquakes

Vegetation protection

Sediment control

Visitors

Appearance

27.   POLAND

Reporter: Andrzej Wita

27.1.   MAIN PRINCIPLES OF DAM SAFETY MANAGEMENT

Poland officially the Republic of Poland is a unitary state with 16 administrative provinces (voivodeships). Poland is a democracy, with a president as a head of state, whose current constitution dates from 1997. The Polish legal system is based on the principle of civil rights, governed by the code of Civil Law.

There are 69 Polish dams in the ICOLD World Register of Dams (42 > 15 m, 18 > 30 m, and 3 > 60 m).

27.1.1. Legal framework for dam safety

There are two national Polish acts covering aspects of dam safety:

- Building Law, July 7, 1994 (Journal of Laws 1994, No. 89, item 414) with amends.

- Water Law, July 18, 2001 (Journal of Laws 2001, No. 115, item 1229) with amends.

Hydraulic structures in Poland are treated as regular building structures and must comprise the specifications of the Building Law. The structures should be designed and built in a manner described in regulations including technical-building and in accordance with the rules of technical knowledge. The building structure should be used appropriately and according to environmental protection requirements and be maintained in a suitable technical and esthetic condition.

The newest amendment of the Water Law of 5 January 2011 changed regulations of the technical and safety conditions control of dams as well as appointed the National Dam Safety Survey defining its tasks.

According to the binding Water Law, I or II class State owned dams are subjected to tests and measurements enabling the evaluation of their technical and safety level conditions, carried out by the National Dam Safety Survey. The same procedure applies to other state-owned dam structures if they are in a bad technical condition that poses a threat or could pose a threat to those structures and the National Water Management Chairman prepares a list of such structures and sends it to the National Dam Safety Survey. New regulations are more precise and unambiguous, disabling different interpretations. The Institute of Meteorology and Water Management plays the role of the National Dam Safety Survey.

State/ Country	Law/Act concerning dam safety	Decrees etc. concerning dam safety
Poland	yes	yes

27.1.2. Responsibilities for dam safety

The dam owners are responsible for the dam safety (from design phase to operation phase/ until decommissioning). Dam owners can be private (persons/companies), municipalities or State.

27.1.3. Arrangements for independent dam safety supervision

The beginnings of technical dam monitoring date back to 1960 in the Institute of Water Management. Then in January 1973 Division for Analysis and Measurements of Hydraulic Structures was launched in Institute of Meteorology and Water Management. It was the first research unit in Poland that initiated comprehensive measurements and studies as well as worked out assessments of technical state and safety of hydraulic structures. Since 1991 it was renamed Dams Monitoring Centre.

Over 200 hydraulic structures are inspected every year by the Centre that carries out measurements and professional inspections, periodic and extraordinary inspections of dams, assessments of technical state and safety as well as prepares expert advice. Verification of measurement systems is also made and data on dams are archived.

Every year a "Report on technical state and safety of hydraulic structures in Poland" is prepared. Until 1995 it covered all annually inspected structures, while in the following years only structures run by units supervised by the Minister of Environment. The Report was drawn up upon request of the concerned minister, and now it is ordered by the Chairman of the National Water Management Board. Technical state and safety of hydraulic structures are described in a synthetic way as well as particular structures included.

According to the binding Water Law, I or II class State owned dams are subjected to tests and measurements enabling the evaluation of their technical and safety level conditions, carried out by the National Dam Safety Survey. The same procedure applies to other state-owned dam structures if they are in a bad technical condition that poses a threat or could pose a threat to those structures and the National Water Management Chairman prepares a list of such structures and sends it to the National Dam Safety Survey. The Institute of Meteorology and Water Management plays the role of the National Dam Safety Survey.

The dam safety has to be under supervision of licensed engineers belonging to the Polish Chamber of Civil Engineers.

27.2. DAM CLASSIFICATION

27.2.1. General principles of classification

Dam classification is defined in the Regulation of Ministry of Environment on the technical conditions which should be met by water management structures and their location of 10 April 2007 (Journal of Laws No. 86, item. 579) [3]. In the table below the classification criteria of permanent dams are presented on the basis of that regulation. There are four classes of dams marked with roman numbers: I, II, III and IV. Class I is the most important. The class of dam depends on:

- height of water level rise

- kind of foundation (rock, soil)

- reservoir capacity

- inundation area as result of dam damage

- population in inundated area

In the table the rules to assign dams to one of the 4 classes were presented and in Fig. 3 the location of I and II class dams in Poland.

Summary table on classification of dams:

Parameter	Class				Remarks
	I	II	III	IV	
Water level rise (1) H [m]	H > 30	15 < H ≤ 30	5 < H ≤ 15	2 < H ≤ 5	Bedrock
	H > 20	10 < H ≤ 20	5 < H ≤ 10	2 < H ≤ 5	Not bedrock
Reservoir capacity V [hm³]	V > 50	20 < V ≤ 50	5 < V ≤ 20	0,2 < V ≤ 5	Capacity at the maximum water level
Area flooded by a dam failure flood wave F [km²]	F > 50	10 < F ≤ 50	1 < F ≤ 10	1 ≤ F	Flooded area water depth over 0.5 m
Population in the dam failure flood L [no. of persons]	L > 300	80 < L ≤ 300	10 < L ≤ 80	10 ≤ L	Apart from regular citizens, offices and factories employees as well as visitors etc.

(1) Water level rise of the hydraulic structure: the difference between absolute height of the maximum water level and the tail water level corresponding to average low water level. In case of foreseen bed erosion or channel erosion it should be taken into account. The structure does not operate during low water the lowest height of the nearest area should be taken into account natural or artificial.

27.3. TECHNICAL FRAMEWORK OVERVIEW

Details and practical solutions to the requirements are given in the technical guideline (edited and issued by Institute of Meteorology and Water Management in 2008) "Guideline for inspection and assessment hydraulic structures". This guideline is recommended to be applied in Poland and is obliged to be used by the Dam Monitoring Center.

27.4. SUPPLEMENTARY INFORMATION

More information about dam safety management, legislation and classification can be found in [1], [2] and [3].

[1] Zaleski J., Wita A., Kosik A.: Dams safety in Poland – regulatory framework, organization structure, the ICOLD Congress, Kyoto, Japan 2012.

[2] Regulation of the Minister of Environment dated 20 April 2007 on technical conditions to be comprised by water management building objects and their location (in Polish).

[3] Guideline for inspection and assessment hydraulic structures, Institute of Meteorology and Water Management, Warsaw, 200 (in Polish).

28.   PORTUGAL

Reporter: José R. Afonso

28.1.   MAIN PRINCIPLES OF DAM SAFETY MANAGEMENT

Portugal, officially the Portuguese Republic, is composed of 18 districts in mainland, and two autonomous regions: the Atlantic archipelagos of the Azores and Madeira.

28.1.1. Legal framework for dam safety

- RSB - Regulamento de Segurança de Barragens (Regulations for the Safety of Dams), Decreto Lei (Decree Law) nº 344/2007 issued in October 15, 2007 (replaces 1990's RSB)

- Normas de Projecto de Barragens (Standards for Dam Design), Portaria (Order or Regulation) 846/93 issued in September 10, 1993

- Normas de Observação e Inspecção de Barragens (Standards for Dam Observation and Inspection), Portaria (Order or Regulation) 847/93 issued in September 10,1993

- Normas de Construção (Standards for Construction), Portaria (Order of Regulation) 246/98 issued in April 21, 1998

- Regime Contra - Ordenacional do RSB (Penalties for the violation of the RSB), Lei (Law) n.º 11/2009, issued in March 25, 2009

- Regulamento de Pequenas Barragens (Regulations for the Safety of Small Dams), Decreto Lei (Decree Law) nº 409/93, issued in December 14, 1993.

28.1.2. Responsibilities for dam safety

The entities concerned with the control of dam safety are:

- Owners, which have overall responsibility for the dams;

- Portuguese Environment Agency (APA), from the Ministry of Environment, that acts as Dam Safety Authority, having general competence in supervising the owner's compliance with the Regulations.

- National Laboratory of Civil Engineering (LNEC), that provides the Authority with technical support for chosen dams of Class I (high potential damage);

- National Authority for Civil Defense, Authority regarding the preparation of emergency plans;

- Dam Safety Commission (CSB), comprising major stakeholders, that analyses the overall Portuguese dam safety progress, at least once a year, reports to the Government and gives its view on owners' complaints about Authority decisions.

28.2. TECHNICAL FRAMEWORK OVERVIEW

The technical framework is comprised in the Legislation. At present the main binding document regarding dam safety is the "Portuguese Regulations for the Safety of Dams" (RSB), first issued as a Decree-Law in 1990 and afterwards revised and re-issued in 2007.

The "Regulations for the Safety of Dams" (RSB) applies to:

    a.   dams with more than 15 meters of height; or

    b.   dams with a reservoir capacity of over 100,000 m³; or

    c.   smaller dams presenting a high potential damage associated to the downstream inundation zone (Class I)

For smaller dams, a "Regulations for Small Dams" apply, which is now under review (should be simplified).

There are also codes of practice for design, construction and observation and inspection of dams (the one for the operation of dams has not yet been published) for due and proper execution of the above-mentioned Regulations. These are all under review.

The mentioned documents are mandatory and constitute legal obligations for all dam owners concerned within the scope of the Regulations.

28.3. DAM CLASSIFICATION

According to RSB dams are categorized according to:

- dimensions:
  - "Large dams": h > 15 m or V > 1 hm³
  - Smaller dams
- associated potential damage, namely downstream
  - Classes I, II, III taking into account loss of life, damage to property and environment, according to the table.

Class	Loss of life, damage to property and environment
I	Number of residents $\geq$ 25
II	Number of residents < 25; or important damage to property; or important environment losses, difficult to recover; or damage to hazardous facilities
III	All the other dams

29.   ROMANIA

Reporter: Iulian Asman

29.1   MAIN PRINCIPLES OF DAM SAFETY MANAGEMENT

Romania is a unitary state divided into 41 counties.

There are more than 2,000 dams in Romania out of which over 200 are large dams according to the ICOLD criteria. The main purposes of large dams are hydropower, flood control and water supply.

29.1.1. Legal Framework for Dam Safety

The main Romanian Dam Safety legislation complementing the Water Law includes the Dam Safety Law, Government Emergency Ordinances and 12 Technical Norms for Hydraulic Works.

The Dam Safety Law stipulate:

- an obligation to provide continuous surveillance

- institution of a "safe operation licensing of the dam" on the basis of a safety assessment of existing dams, carried out by technical experts approved by the Ministry of Public Works and certified by the Ministry of Environment and Forests (MEF)

- institution of sanctions and fines for dam owners that endanger the safety of the population, goods and environment.

The main Romanian regulations regarding dam safety and related items are:

- Water Law (Law no. 107/8.10.1996) with subsequent modifications and complements;

- Dam Safety Law (Law no. 466/18.07.2001) for the approval of Emergency Government Ordinance no. 244/ 28.11.2000 concerning Dam Safety;

- 12 Technical Norms for Hydraulic Works published in the period 2002–2003 for the application of Law no. 466; they regard dam surveillance, safety evaluation reports, constitution of corps of experts, etc.

- Small Reservoirs Law (Law no. 13/ 11.01.2006) for the approval of Emergency Government Ordinance no. 138/29.09.2005 regarding the Safe Operation of small reservoirs such as fishing, recreation or reservoir of local importance.

State/ Federal State/ Province/Territory	Law/Act concerning dam safety	Decrees etc. concerning dam safety
Romania	yes	yes

29.1.2. Responsibilities for Dam Safety

Dam owners are directly responsible for accomplishing and keeping the safety of their dams in conformity with the Dam Safety Law.

All dams must have surveillance programmes and must be submitted at certain time intervals to an overall safety assessment performed by certified experts according to a specific norm: NTLH – 022 for water retaining structures and NTLH – 023 for tailing dams.

The contents of a typical dam safety assessment report include data concerning the design, construction and operation of the facility, a report drawn up following the technical inspection, results of complementary investigations (if needed), a report regarding technical, functional and safety in operation conditions of the facility and in the final part proposals and recommendations regarding the terms for continuing operation. The dam safety assessment reports have to be submitted to the responsible Dam Safety Evaluation Commission (see below) to get a "safe operation licensing" for the dam.

29.1.3. Arrangements for Independent Dam Safety Supervision

The governmental supervision of dam safety is carried out by the National Commission on Dam Safety, CONSIB. CONSIB organizes and supervises the activity of five Dam Safety Evaluation Commissions, DSEC:

- Dams rated into the highest categories A and B (see below) are managed by one central DSEC, which is organized by Ministry of Environment and Forests.

- Dams rated into categories C and D (see below) are managed on a territorial basis by four DSECs organized by the Romanian Water Authority.

DSEC are dealing with overall dam safety reports of all the dams, regardless of their category of importance, covering larger periods of time and more general aspects besides the monitoring aspects. The content of this kind of reports is detailed in the NTLH – 022 for dams and NTLH – 023 for tailing dams.

In accordance with the Dam Safety Law, the dam monitoring (surveillance) activity is organized on three levels:

- Level I where collection and primary interpretation of data takes place;

- Level II comprising the synthesis of visual observations, measurements and technical inspections gathered in periodic (annual) monitoring reports;

- Level III including the analysis and approval of the monitoring reports by the Dam Monitoring Commission, DMC. The major owners (ex: Romanian Water Authority, Hidroelectrica – the Hydropower company, own almost all the water retaining structures rated into A and B) have such a commission as a consulting body comprised of experts and specialists for their A and B dams. The other owners must affiliate their important dams (in general A and B or those dams that need a particular attention from the monitoring point of view according to specialists) to one of the existing Commissions. The dam safety surveillance activities for category C and D dams are organized by the dam owners themselves with the assistance of the water directorates of the Romanian Water Authority at the local level.

It is the task of CONSIB to approve the structure of DMC and of the five Dam Safety Evaluation Commissions and their annual activity reports.

29.2. DAM CLASSIFICATION

29.2.1. General Principles of Classification

In the Dam Safety Law dams are defined as water retaining structures (water filled and dry), tailing dams and special hydraulic structures (hydro power plants and navigation locks within the damming structure, open diversion works executed in backfill, penstocks, water tanks over 5,000 m^3). The classifications refer to all water retaining dams. For the tailing dams there is a similar type of classification.

There are two basic classifications of dams:

- Classification of importance regarding hydrological safety for selection of design flood (the first chronologically)

- Classification regarding the risk associated with the dam used for all types of dams. This classification refers to the categories of importance regulated by law no. 466 and the pertaining technical norms NTLH

29.2.2. Dam Classes Overview

Class of importance regarding hydrological safety, I-IV

The criteria for the selection of design flood are based on dam rating into four classes of importance, according to STAS 4273 – 83, with demands for greater hydrological safety for the dams rated in the upper classes of importance. The classification is based on:

- the dam dimensions

- reservoir volume

- reservoir benefits (installed hydropower, water supply discharge, irrigated area, etc)

- the dam failure consequence

Categories of importance depending on the dam associated risk, A-D

The Dam Safety Law and the Technical Norm NTLH – 021 stipulate that all water retaining dams shall be rated into four categories of importance in terms of their associated risk. The Dam Associated Risk (DAR) is defined in terms of three indices regarding:

- dam characteristics and conditions of the site

- dam behaviour and condition of the structure

- dam failure consequences

29.3. TECHNICAL FRAMEWORK OVERVIEW

Besides the norms pertaining to Dam Safety Law, NTLH, there are also other norms and guidelines issued by the Ministry of Public Works and Land Development included in the Technical Regulations concerning the design and execution of hydraulic structures and river development and improvement. Below some of those related to dam safety and monitoring are listed.

Index Technical Regulation	Title of the Technical Regulation
NP 076–2002	Norm for the design, execution and evaluation of safety at seismic loads of hydraulic structures in the damming construction elaborated by UTCB – Romanian for Technical University of Civil Engineering Bucharest
NP 087–2003	Norm for surveillance of the hydraulic structure behaviour elaborated by ISPH - Romanian for Institute for Hydropower Studies and Design
NP 090–2003	Norm for the seismic instrumentation of dams elaborated by UTCB – Romanian for Technical University of Civil Engineering Bucharest

29.4. REFERENCES

[1]. Dam Safety Law (no. 466) and technical norms for implementation (NTLH), Bucharest 2001–2003 (in Romanian/English)

[2]. Stematiu D. 2000 Dam Safety. Chapter 9 of the volume "Dams in Romania" ROCOLD, Bucharest (in English)

[3]. Stematiu D., Ionescu St., Marinescu P. 2001 Progress in Legislation based on risk management. Procedures of ICOLD European Symposium on Dams in European Context, Geiranger (in English)

30.   RUSSIA

Reporter: Evgenyi Bellendir

30.1.   MAIN PRINCIPLES OF DAM SAFETY MANAGEMENT

30.1.1. Legal framework for dam safety

Russia has a comprehensive dam safety legislation system consisting of the National Law "On Safety of Dam Structures" supported by lower level laws, regulations and national standards defining (1) requirements of safety declaration of dam structures, (2) classification of natural and technical hazards, (3) determination of the financial and civil liability in case of dam failure, (4) assessment of the loss of life and impact to humans and properties, (5) methodology for preparation of the safety declaration etc.

State/ Federal State/ Province/Territory	Law/Act concerning dam safety	Decrees etc. concerning dam safety
Russian Federation	yes	yes

30.1.2. Responsibilities for dam safety

The dam owner or operator is responsible for dam safety.

The state supervision of dam safety is performed by Rostekhnadzor, the federal Service on environmental protection and nuclear power. Rostekhnadzor can perform unplanned inspections of dam safety, participate in regular commissions, design expertise, give permissions for construction, operation and decommissioning of hydraulic facilities, develop standards on dam safety and safety management programs, etc.

30.1.3. Arrangements for independent dam safety supervision

The national level Commissions including representatives of the state surveillance and controlling authorities, dam designers and scientific institutes are involved in supervision of dam safety. They perform regular inspections of dams under operation and the dam owners have to perform measures to improve dam safety following the requirements issued by commissions based on the inspection results.

30.2.   Dam classification

30.2.1. General principles of classification

Dams are classified according to the requirements stated in the National Law "On safety of dam structures" and national standard "SNiP 33-01-2003". Dams are classified according to the dam height and the consequences in case of dam failure.

30.2.2. Dam classes overview

According to the requirements of Rostekhnadzor dams are classified by the safety level as follows:

I	The normal safety level: the dam is operated in accordance with the design, standards and rules currently in force, the values of safety criteria are within permissible limits, the dam is operated in accordance with current legislation, regulations, standards and rules, following the requirements of supervisors.
II	The lowered safety level: measures on dam safety are not performed or directions of state supervisory bodies are not fulfilled or other violations of operation rules.
III	The unsatisfactory safety level: lowered mechanical or seepage strength, safety criteria are not fulfilled, other deviations from the design which may cause dam failure.
IV	The hazardous safety level: dam operation is under the risk due to decreased strength and stability of dam structures and foundation, measured safety parameters are out of normal range of dam operation.

Number of classes: 4.

30.3.   TECHNICAL FRAMEWORK OVERVIEW

In accordance with the requirements of the National Law "On technical regulation", the system of technical standards in the field of dam safety including technical regulations, national, branch and enterprise standards is currently being developed.

The system of state construction codes and rules (SNiP) is in force with the same status as a law. In addition, there is a system of national standards (GOST) used for determination of parameters of all types of construction materials and soils, as well as some technical regulations, recommendations and manuals for design of dams and foundations.

31.   SCOTLAND

Reporter: James Ashworth

31.1.   MAIN PRINCIPLES OF DAM SAFETY MANAGEMENT

Great Britain comprises the territory of England, Scotland and Wales. England, Scotland and Wales forms the United Kingdom along with Northern Ireland. UK is a constitutional monarchy and a unitary state. England and Wales are under English Law, while Scotland has its own Scots law.

The dam safety authority in Scotland is currently the 32 Local Authorities and they regulate under the Reservoirs Act 1975, but under the new reservoir safety legislation, the Reservoirs (Scotland) Act 2011 this will be changing and the Scottish Environment Protection Agency (SEPA) will take on the role of regulator. The following facts are valid for Scotland only.

There are approximately 660 reservoirs in Scotland that fall under the Reservoirs Act 1975, but this number is expected to increase significantly with the implementation of the Reservoirs (Scotland) Act 2011 which lowers the registration threshold from 25,000 cubic metres to 10,000 cubic metres.

31.1.1. Legal framework for dam safety

Requirements regarding dam safety are given in the Reservoirs Act 1975, HMSO, 1975 and the Reservoirs (Scotland) Act 2011 – passed through the Scottish Parliament but yet to be commenced.

State/ Country	Law/Act concerning dam safety	Decrees etc. concerning dam safety
Great Britain (Scotland)	yes	yes

31.1.2. Responsibilities for dam safety

- The dam owners are entirely responsible for the dam safety.

- Enforcement only by Government (Local Authorities at present but passing to SEPA).

- Inspecting Engineers and Supervising Engineers from a Panel of Engineers.

31.1.3. Arrangements for independent dam safety supervision

Currently under the Reservoirs Act 1975 any dam greater than 25,000m³ above the natural ground has to be inspected at least every 10 years by an Inspecting Engineer. Panel Engineers inspect against guidelines and recommend works if required. Recommendations in the interests of safety have to be completed and can be enforced.

All registered reservoirs should be under the supervision of a Supervising Engineer who undertakes regular monitoring of the dam. Supervision by a trained engineer (Supervising Engineer) involves a visit at least once a year.

The Reservoirs (Scotland) Act will move towards a risk-based approach which will change the above approach.

31.2. DAM CLASSIFICATION

31.2.1. General principles of classification

Dams are classified according to consequence of failure. Criteria for classification: life safety (loss of life or population at risk) and economic losses. Dam classification is moving to consequence categorization and risk rather than just on retained volume. Classification is equal for all dams/ reservoirs (not type dependent).

Number of classes: 4

31.2.2. Dam classes overview

Currently dams are classified via guidelines associated with the Reservoirs Act, 1975 according to consequence of failure.

A	10 or more killed
B	0–10 or more killed
C	no one killed but economic loss
D	no losses

Under the new legislation the risk designations will be 'High', 'Medium' or 'Low'. These designations will be dependent on the consequences of an uncontrolled release of water on a number of receptors. These receptors include human health, infrastructure, power supplies, economic interests, the environment and cultural heritage.

31.3. TECHNICAL FRAMEWORK OVERVIEW

There are guidelines on floods and seismicity:

- Floods: An Engineering Guide & Reservoir Safety: ICE 1992.

- An Engineering Guide to Seismic Risk to Dams in the United Kingdom – BRE 1991.

31.4. SUPPLEMENTARY INFORMATION

More information about dam safety management, legislation and classification can be found on the BDS (British Dam Society) web site and numerous papers from international journals and conference proceedings.

SEPA also has a web site on which information regarding the Reservoirs (Scotland) Act 2011 can be found.

32.   SERBIA

Reporter: Ignjat Tucovic

32.1.   MAIN PRINCIPLES OF DAM SAFETY MANAGEMENT

Republic of Serbia is located in Southeast Europe and arose from former Yugoslavia. Serbia is a parliamentary republic with a multi-party system with a unicameral National Assembly. The executive authority is exercised by the prime minister and cabinet members. Apart the government there is also a president with no executive, legislative, or judicial authority.

Serbia is organized into 5 distinct regions, further divided into 29 districts. Belgrade is a separate territorial unit established by the Constitution and law, while Šumadija and Western Serbia and Southern and Eastern Serbia are directly subordinated to country authorities.

There are over 63 large dams (second to the information on the Register of dams) in Serbia and 60 of them are higher than 15 m. The first larger dams were constructed in 19th Century, while in the beginning on 20th Century the first dam with purpose of hydro power production has been constructed. Most of the dams for the hydro power production have been constructed between 1950 and 1980 while most of the dams for water supply have been built after 1960. In the last decades, the construction of dams has decreased significantly.

32.1.1. Legal framework for dam safety

Serbia's legislation system concerning dam safety is mostly based on three major pillars:

- Surveillance and monitoring of dams – the law brings up the official classification

- Water act – concerning maintenance and operation of hydro structures

- Reports and analysis of monitoring and procedures during the construction as well as soil and structure behavior during the exploitation period

The main Serbian regulations concerning the dam safety and related items include:

- Former Yugoslav monitoring and surveillance regulations; Former Yugoslav regulations Official Gazette SFRY No.7 of 16. February 1966.

- System of monitoring and communication act; Official Gazette of 16. February 1990.

- Water act; Official Gazette of the Republic of Serbia No. 46 of 1991, completion 53/1993, 67/1993, 48/1994 and 54/1996.

- Regulations on soils and structures report contents and monitoring procedures during the construction and exploitation; Official Gazette of the Republic of Serbia No. 13, of 15. April 1998.

- Environment protection act; Official Gazette of the Republic of Serbia No. 135, 2004.

State/ Country	Law/Act concerning dam safety	Decrees etc. concerning dam safety
Serbia	yes	yes

31.1.2. Responsibilities for dam safety

The dams in Serbia are owned by state. The management and operation is entrusted to different public companies (The hydro power companies and Water management companies). These public companies coordinate the management and operation of the dams with maintenance and monitoring procedures, which are organized partially by different sectors of the same company and partially by specialized consulting companies. The one responsible for dam safety is the dam owner by the law.

31.1.3. Arrangements for independent dam safety supervision

The dam safety evaluation lies within monitoring and surveillance of dams – the performance of it is compulsory by law. The operation, maintenance and monitoring of dams is compulsory by law and is organized and performed by public companies (see above) continuously in accordance to Operation and maintenance instructions and Monitoring instructions prepared by consulting companies according to legislation already at the design stage. This Operation and maintenance instructions define the extent of surveillance and monitoring works (time schedule, quantity and quality of measurements). The monitoring and surveillance results are summarized and analyzed in annual reports which are obligatory by law for owner of a dam (meaning the one who manages it), but can be done by the company that the dam is entrusted to. Apart from the results of the analysis and findings the reports must contain also the assessment of dam safety.

The review of collected data and the elaboration of annual reports is entrusted (by law) to the consulting companies which designed the dam. When preparing the annual or periodical reports, official regulation, worldwide recognized good practice and ICOLD recommendations are followed. The reports are reviewed at the end by the state supervising institution Civil engineering inspectorate. If necessary, the Inspectorate orders the dam owner to carry out relevant actions to improve dam safety.

The re-evaluation of dam safety is preformed periodically once each 10 years (using modern techniques and criteria of the dam design). The activity includes analysis of the potential risk of dam failure (with relevant consequences) as well as the control of crucial safety elements such as stability, seepage, stress patterns, deformation on dam surface and dam structure etc.). The re-evaluation is obligatory by law.

ANNUAL SCHEME   EVERY TEN YEARS

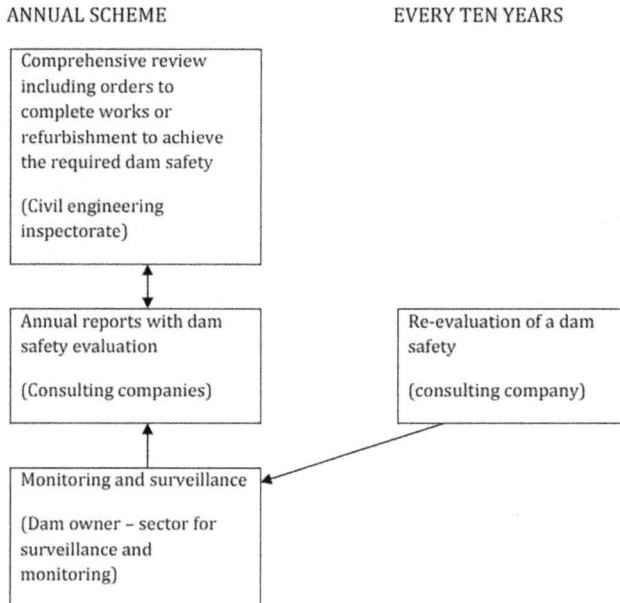

32.2. DAM CLASSIFICATION

32.2.1. General principles of classification

Dam classification is defined by the Former Yugoslav monitoring and surveillance regulations and it does not take risk assessment into consideration. However, a Methodology for hazard potential classification of existing dams has been prepared by the National committee on dam safety in 1997, but until now it has not been made official.

32.2.2. Dam classes overview

The dams are divided in categories in accordance to existing ICOLD classification:

Large dams		Other dams	Tailing dams
H> 15 m		All the dams and water retaining structures which don't apply with the listed conditions for large dams	Specially treated*
H > 10 m	Length of the crest L_c>500m		
and:	Volume of reservoir V>100,000m³		
	Spillway capacity Q_{sc}>2,000m³/s		
*) Tailings dams are treated as a special type of dams, but they are practically within the general dam categorization.			

In the unofficial – proposed – method dams are divided in categories according to the hazard potential and the behavior index.

Concerning the Hazard Potential (HP) dams are classified as with low, moderate or high HP, and as far as the Behavior Index is concerned the state of the dam is described as very good, good or unsatisfied. These two factors are combined to define the Dam Safety Index. Hazard Potential contributes 40% and registered dam behavior 60%. Categorization by the Hazard Potential is based upon:

- project importance;

- life and property losses;

- dam type;

- type of evacuators;

- design floods;

- seismic hazard;

- conformity of basic and current design criteria;

- reservoir management system

32.3. TECHNICAL FRAMEWORK

The standards and guidelines exist. The process of evaluation of dam safety is based on standards concerning the design of different kind of structures in general:

- EN 1996 - Eurocode 6: Design of Masonry Structures.

- EN 1977 - Eurocode 7: Design of Concrete Structures.

- EN 1997 - Eurocode 7: Geotechnical Design.

- EN 1998 - Eurocode 8: Design Provisions for Earthquake Resistance Structures.

- ICS 91.100.30 - JUS U.E3.010 - Yugoslav standard: Hydrotechnical concrete - Technical requirements for manufacture and use. Issued 1987.

- ICS 91.100.30 - JUS U.M1.014 - Yugoslav standard: Concrete - Influence of aggressive materials and protection against them.

- ICS 91.100.30 - JUS U.M1.015 - Yugoslav standard: Concrete- Concrete, hardened - Determination of the depth of penetration of water under pressure.

- Guidelines defining consequences of failure or overtopping of large dams issued by the former Yugoslav Federal Committee for Agriculture, January 1975.

32.4. SUPPLEMENTARY INFORMATION

More information about dam safety management, legislation and classification can be found in:

[1] HTTP://WWW.ICID.ORG/TF5_PAPER.PDF
[2] HTTP://EN.WIKIPEDIA.ORG/WIKI/SERBIA

33. SLOVAKIA

Reporter: Peter Panenka

33.1. MAIN PRINCIPLES OF DAM SAFETY MANAGEMENT

Slovak Republic is a parliamentary democracy managed by unicameral parliament, government and president. The state legislation is unified. The matter on dam safety is managed by Ministry of Environment and on lower level by district offices of the environment at the county (from 2013).

There are approximately 950 dams and other water structures (dikes, tailings dams etc.). 50 dams are registered in the ICOLD Register of Large Dams (5 concrete dams and 45 earth-fill or rock-fill dams). There are 550 small dams, 130 levees for flood protection and 61 tailing dams are situated in Slovakia. The number of dams is increasing due to ongoing registration of existing dams and planning and construction of new (especially small) hydro power projects.

The main purposes of large dams are hydropower production, drinking water supply, flood protection and water accumulation. Historical small dams serve mostly for fishing and recreation.

33.1.1. Legal framework for dam safety

- Act No. 364/2004 - "Water Act"

- Decree of the Ministry of the Environment No. 458/2005 establishing details of the dam safety surveillance and supervision on hydraulic structures

- Act No. 7/2010 - "Flood Protection"

- Decree of the Ministry of the Interior No. 388/2006 – warning system for endangered inhabitants downstream.

State	Law/Act concerning dam safety	Decrees etc. concerning dam safety
Slovakia	yes	yes

33.1.2. Responsibilities for dam safety

The owner or administrator of a dam is responsible for the dam safety. They are obliged to keep the dam safe, in good condition, and perform observations and measurements (automated and manually) with a frequency according to the dam category and the surveillance program. The results are analyzed and worked on in annual reports (I. cat.).

Dam safety surveillance begins already in the planning stage of the dam. At that time the project of measurements and observations, and the dam surveillance program, are prepared. Then the surveillance process continues during dam construction and other periods of the dam existence – the verifying period and during permanent dam operation.

Dam safety is controlled by Dam Safety Surveillance Department (DSSD) which is a part of Vodohospodárska výstavba (Water Management Construction) state company. This state enterprise is entrusted by Minister of Environment of Slovak Republic to perform dam surveillance on all Slovak dams of I. and II. category. DSSD also provides categorization of all water constructions (dams, flood protection dikes, tailings dams, etc.), keeps the list of all categorised hydraulic structures and prepares reports about dam surveillance results for Ministry of Environment and also for regional office of environment and for dam owners. Dam safety surveillance is executed by means of measurements, analysis, tests, recognition and recommendation and results are evaluated with the goal to achieve needed dam safety level.

For each dam of the categories I and II a "main supervisor" is designated – an employee of Vodohospodárska výstavba s.e., who is a certified person for dam safety surveillance. The main supervisor for the dam categories III and IV could be any person with DSS certification.

33.1.3. Arrangements for Independent Dam Safety Supervision

The Ministry of Environment inspects the performance of dam safety surveillance on the dams of the highest categories - I (every year) and II (every second year) by means of local water authorities (District Offices of the Environment). The results of the surveillance on the other water structures - categories III and IV - are elaborated once in four years or once in ten years.

Inspections of the dam safety surveillance performance are executed by local hydraulic authorities (regional, district). Inspection includes checking of regular reports, visual and functional control in-situ, and in when necessary the prescription of preventive measures for improving dam safety or upgrading.

The owner can be penalized if dam surveillance is not provided or if prescriptions are not accepted.

33.2. TECHNICAL FRAMEWORK OVERVIEW

The basic guideline is the Decree of Ministry of Environment about realization of Dam Safety Surveillance on hydraulic structures (Collection of Laws 2005, No. 458, September 12, 2005).

There are many technical standards and guidelines on the subject of dam safety surveillance, although they are written not only for dam safety but for water management, stability of construction and buildings etc. A selection of them is:

- STN 73 0035 Loading of building structures

- STN 73 0036 Seismic load of buildings

- STN 73 0090 Geological exploration for construction purposes

- STN 73 0405 Measurement of displacement and deformation of construction objects and their parts

- STN 73 2005 Injection works in construction

- STN 73 2030 Load testing of structures

- STN 73 2400 Construction and the building inspection of concrete structures

- STN 73 3050 Earthworks

- STN 73 6500 The calculation of wave impacts

- STN 73 6503 Load of water works by water pressure

- STN 73 6506 Load of water works by ice

- STN 73 6510 Basic water management terminology

- STN 73 6512 Naming a brand of water management – modifications of flows, rivers

- STN 73 6516 Dams

- STN 73 6517 Hydropower

- STN 73 6805 Hydrological data of surface water

- STN 73 6814 Design of dams

- STN 73 6822 Crossing and overlapping management and communications with watercourses

- STN 73 6824 Small water reservoirs

- STN 73 6824 Low earthfill dams

- STN 73 6850 Earthfill dams

33.3.   DAM CLASSIFICATION

33.3.1. General principles of classification

The criteria of classification are described more in detail in an Ordinance of the Ministry of Environment.

The class of a dam is based on the assessment of a risk factor "F", which results from the danger posed by the existence of the dam. This classification is based on a point evaluation of the damage at the dam structure itself, human live losses and other damages downstream (buildings, area, industry, agriculture, infrastructure, environment etc.):

Category I	$F \geq 1,000$ points
Category II	$150 \leq F < 1\,000$
Category III	$15 \leq F < 150$
Category IV	$F < 15$ points
(The value of 1 classification point is approximately 116,700 €)	

Examples of parameters that are taken into consideration are the height of damming, the type of dam structure, the reservoir capacity, the downstream area morphology, the number and density of population living in the inundated area, infrastructure, industry, environmental or historic aspects etc.

The dam classification does not depend on the different types (and uses) of reservoirs. Therefore, tailings dams, dikes (levees) and other similar damming structures are also classified in the same way as dams.

33.4. REFERENCES

[1] Act No. 364/2004 - "Water Act"

[2] Decree of the Ministry of the Environment No. 458/2005 establishing the details of the dam safety surveillance and supervision

[3] Act No. 7/2010 - "Floods Protection"

[4] Decree of the Ministry of Interior No. 388/2006 – about construction of warning system for inhabitants living and endangered downstream

34.   SLOVENIA

Reporter: Nina Humar

34.1.   MAIN PRINCIPLES OF DAM SAFETY MANAGEMENT

Republic of Slovenia is a unitary state and a parliamentary democracy situated in South-Central Europe and arose from former Yugoslavia. The head of state is the president, who has mainly a representative role. The executive and administrative authority in Slovenia is held by the Government of Slovenia headed by the Prime Minister and the council of ministers or cabinet, who are elected by the National Assembly. The legislative authority is held by the bicameral Parliament of Slovenia, characterized by an asymmetric duality.

Slovenia is subdivided into 211 municipalities - bodies of local autonomy – governed by mayors - which are grouped into 62 administrative districts "Administrative Units". Administrative units are territorial sub-units of government administration.

There are over 38 large dams second to the actual ICOLD categorization in Slovenia and 35 of them are higher than 15m. The first dam was constructed in 1769, but the greater part of the dams was constructed between 1950 and 1990. The first dams for hydro-power production were constructed in the mid-thirties of the 20th century. Between 1960 and 1996 most of the existing dams for irrigation and water protection were constructed. In the last decades the construction of dams has turned again to construction of dams for hydropower production.

A big part of the large dams in Slovenia are state owned - only few are owned by private companies or municipalities. The large dams are under the supervision of the Ministry of agriculture and environment (2012) and the Inspectorate of environment. The emergency preparedness plans are under the supervision of the Ministry of defense. Smaller and medium size dams are not explicitly subject to Inspectorate's supervision.

34.1.1. Legal framework for dam safety

The basics regulations concerning dam safety and safety of water constructions are:

- Construction act (Official gazette RS No.110/2002, and completion 2010)

- Water act (Official Gazette RS No.67/2002 and completion in 2008)

- Spatial planning act (Official Gazette RS No.33/2007 and completion 2009)

- Protection Against Natural and Other Disasters Act (Official Gazette RS No.51/200 and completion 2010)

- Regulation for classification of very demanding, demanding and simple engineering structures, about the conditions for construction of simple engineering structures that do not need building permit and about the type of construction works that are in reference with structures and appurtenant land (Official Gazette RS No.114/ 2003 and completion)

- Regulation on the classification of types of structures and facilities of national importance (Official Gazette RS No.119/2011)

- Former Yugoslav monitoring and surveillance regulations (Official Gazette SFRY No.7/ 1966)

- Regulations for seismic monitoring of large dams (Official Gazette RS No.92/1999, completion in 2003)

Apart from the regulations mentioned above there are several other laws that influence directly or indirectly the safety of dams:

- Environment protection act (Official Gazette RS No.2004, completion: 2006 and 2008)

- Protection against natural and other disasters act (Official gazette RS No.1994, completion: 2006, 2008, 2010)

- Guidelines on the elaboration of contingency plans (Official Gazette RS No.2002, completion 2008 and 2010)

- Law on Construction Products (Official Gazette No.52/2000, completion)

State/Country	Law/Act concerning dam safety	Decrees etc. concerning dam safety
Slovenia	yes	yes

34.1.2. Responsibilities for dam safety

The dams are mostly state owned (only few are owned by private companies or municipalities). According to the law (Construction act and Water act) the owner is responsible for dam safety, but in most cases the management and operation is entrusted to different public and semi-private companies (Hydro power companies and Water management companies). These companies take care of the operation and maintenance of the dams as well as the performance of monitoring. These procedures and smaller refurbishment works are normally performed by different sectors of the same company (visual inspections, equipment tests etc.) and in part by specialized consulting companies (geotechnical inspections, specialized controls of the equipment etc.) – coordination of these inspections is organized by the company responsible for management of a single dam.

The operation, maintenance and monitoring of dams is organized and performed by the companies who manages a dam continuously in accordance to Regulation for operation and maintenance of a dam, which is compulsory by the Construction act. The regulation summarizes the requirements defined in Operation and maintenance projects and Monitoring projects prepared by consulting companies (according to legislation) already at the design stage. This project normally defines the extent of the maintenance, surveillance and monitoring as well as time schedules for operation, maintenance and monitoring. These projects are obligatory by law.

The extent of monitoring programs is not specifically defined by the law (however the Former Yugoslav monitoring and surveillance regulation is still, or is again, in force). The manager of the dam is obliged to keep a diary in which all the changes and notifications must be quoted. The annual reports and safety evaluations are not obligatory by law for now, but an effort is put in to tighten the requirements and improve the legislation.

For each large dam (second to the old ICOLD categorization) emergency action plans should be prepared. The emergency action plans are prepared by the company managing the dam – by the engineer in charge for the dam. The existence as well as the application and organization of these emergency procedures are regularly controlled by the Inspectorate of the Ministry of defense.

34.1.3. Arrangements for independent dam safety supervision

The dam safety concept could be divided in three areas/phases of dam safety care:

- structural safety (design and construction phase)

- surveillance and maintenance (operation, refurbishment)

- emergency action planning and alerting of downstream population (extreme conditions)

The improvement of structural safety is to be achieved by control of all aspects of the design phase (project, studies, analysis). The basis for this phase is to be found in Construction act, Spatial planning act and Regulation for classification of very demanding, demanding and simple engineering structures, about the conditions for construction of simple engineering structures that do not need building permit and about the type of construction works that are in reference with structures and appurtenant land. The projects are revised by official auditors and should meet the requirements defined in the regulation mentioned above.

The construction of large dams is supervised by a team of different expert engineers. The additional external supervision is or can be performed by the Inspectorate of civil engineering - the inspectors control the application of principles defined in the laws mentioned above and fulfillment of the requirements defined in project and standards.

The supervision of the performance of a dam (monitoring, operation) is performed by an appointed engineer (at the company who manages the dam). The external supervision is or can be performed by the Inspectorate of Ministry of agriculture and environment, but the inspectorate just controls the application of procedures and activities defined in instruction for operation, maintenance and monitoring. There is no concrete supervision and reviews of regular and periodical reports. Reassessments are not obligatory by law and therefore they are rarely performed. Apart from this there is a slight conflict of interest since the state (Ministry) appears in both roles – the role of the owner and the role of the supervisor.

The existence as well as the application and organization of Emergency action plans and emergency procedures are regularly controlled by the Inspectorate of the Ministry of defense.

34.2. DAM CLASSIFICATION

34.2.1. General principles of classification

Dam classification defined by regulation doesn't take risk assessment into consideration. However, a Methodology for hazard potential classification of existing dams has been prepared by the Faculty of civil and geodetic engineering, but for now it has not been made official.

There are three official categorizations of dams. The first categorization is brought up by Regulation on the classification of types of structures and facilities of national importance (Official Gazette RS No.119/2011) and it applies to dams in design phase - more specifically to the location of the dam and the extent of necessary preliminary research and studies.

Dams		
Very demanding structures		**Demanding structures**
H> 10m	with crest L>50m	All dams and water retaining structures which don't apply with the listed conditions for large dams
H> 4m	with storage capacity exceeding 100,000 m³	
	with exceptionally difficult or problematic foundation conditions	
	which can threaten populated areas, Important public transport ways, utilities and energy infrastructure or good ecological status of areas under the dam	
Dykes		
Very demanding structures		**Demanding structures**
H> 10m	with crest L>50m	All the dams and water retaining structures which don't apply with the listed conditions for large dams
H> 4m	with storage capacity exceeding 100,000 m³	
	with exceptionally difficult or problematic foundation conditions	
H>2m	when failure can threaten populated areas, important public transport ways, utilities and energy infrastructure or good ecological status of areas under the dam	

The second categorization is defined by Regulation for classification of very demanding, demanding and simple engineering structures, about the conditions for construction of simple engineering structures that do not need building permit and about the type of construction works that are in reference with structures and appurtenant land (Official Gazette RS No.114/ 2003 and completion) refers to extent of elaboration in design phase – specifically to accuracy of elaboration and requirements for the design documentation and projects.

Very demanding structures		**Demanding structures**
H> 10 m	For earth fill dams for water or debris flow containment with crest L>250m	All the dams and water retaining structures which don't apply with the listed conditions for large dams
	Concrete dams with crest L>50m	
	All dams with crest L>300 m	

The third classification is presented by The Former Yugoslav monitoring and surveillance regulations – the classification is taken after the old ICOLD categorization but is also taken into account by Protection against natural and other disasters act. It refers to the operational phase and to emergency preparedness:

Large dams		**Other dams**
H> 15 m		All the dams and water retaining structures which don't apply with the listed conditions for large dams
H > 10 m and:	Length of the crest L_c>500m	
	Volume of reservoir V>1,000,000 m3	
	Spillway capacity Qsc>2,000m3/s	
	Special foundation conditions	
	Special design	

NOTE: The unofficial classification based on hazard potential proposed by Faculty of civil and geodetic engineering, University of Ljubljana in 1996. The risk classification is still not installed - there is an initiative to enforce also a classification according to risk, which would categorize the dams into 3 categories:

Hazard potential
Low
Moderate
High

The classification criteria are based on:

- Life loss/population at risk

- Propriety and direct economic loss of third party

- Other socio-economic impact

- Cultural and natural heritage loss

34.3. TECHNICAL FRAMEWORK

The technical framework concerning dam safety is mostly based on standards concerning the design of different kind of structures in general and guidelines for preparation of contingency actionplans.

- SIST EN 1992 - Eurocode 1: Design of Concrete Structures.

- SIST EN 1996 - Eurocode 6: Design of Masonry Structures.

- SIST EN 1997-1-1: Eurocode 7: Geotechnical design - Part 1.

- SIST EN 1998-1: Eurocode 8: Design Provisions for Earthquake Resistance Structures.

And in standards which define the basic requirements about the quality of work and materials used in construction works and refurbishment of dams.

- SIST EN 13383 Armourstone - Part 1: Specification.

- SIST EN 13251 Geotextiles and geotextile-related products - Characteristics required for use in earthworks, foundations and retaining structures.

- SIST EN 13252 Geotextiles and geotextile-related products - Characteristics required for use in drainage systems.

- SIST EN 13254 - Geotextiles and geotextile-related products - Characteristics required for use in the construction of reservoirs and dams.

- SIST EN 13361 Geosynthetic barriers - Characteristics required for use in the construction of reservoirs and dams.

- Guidelines on the elaboration of contingency plans (Official Gazette RS 2002, completion: 2008).

34.4.   SUPPLEMENTARY INFORMATION

More information about dam safety management, legislation and classification can be found in:

[1] HTTP://www.slocold.si

[2] HTTP://EN.WIKIPEDIA.ORG/WIKI/SLOVENIA

[3] HTTP://www.uradni-list.si (Official Gazzette)

[4] Humar, N., Kryžanowski, A. – Dam safety in Slovenia; HydroVision - hydroconference and exhibition, July 19–22, 2011, Sacramento, California, USA

[5] Humar, N., Kryžanowski, A. - Dam safety practice in Slovenia - Dam safety – New challanges, Bezpieczenstwo zapor - nowe wyzwania, Warszawa, Instytut meteorologii i gospodarki wodnej panstwowy, 2011

35. SOUTH AFRICA

Reporter: Chris Oosthuizen

35.1. MAIN PRINCIPLES OF DAM SAFETY MANAGEMENT

South Africa is a parliamentary republic divided into 9 provinces (since 1994). The dam safety authority is The Department of Water Affairs (DWA).

35.1.1. Legal framework for dam safety

Sections 117 to 123 (Chapter 12) of the National Water Act, 1998 (Act No. 36 of 1998), presently enforce dam safety in South Africa.

The Dam Safety Office of DWA implements and administers the Dam Safety Regulations which is in force under the National Water Act (Act 36 of 1998), reference www.dwa.gov.za/DSO.

There are no specific dam safety norms and standards. The Approved Professional Persons (engineers), APP, has the responsibility to determine appropriate standards for a particular dam and the legislation provides for a review of such standards by the Director-General. The appropriate norms are considered to be current acceptable dam engineering practice for site-specific conditions.

State/Provinces	Law/Act concerning dam safety	Decrees etc. concerning dam safety
South Africa (all provinces)	yes	Yes[1]
[1] no norms and standards, only regulations		

35.1.2. Responsibilities for dam safety

The dam owners have to comply with the requirements of the Act and the Regulations and are responsible for registration and classification of their dams (application to DWA).

35.1.3. Arrangements for independent dam safety supervision

The Dam Safety Office of DWA is established to enforce the Regulations.

35.2. DAM CLASSIFICATION

35.2.1. General principles of classification

Criteria for classification: The category classification is determined in accordance with criteria given in the Regulations. It is based on the size class of a dam and its hazard potential rating. Size class is determined by the maximum wall height as shown in Table 1. Hazard potential is based on an assessment of the potential loss of life (PLL) and potential economic loss (PEL) that may result from failure of a dam. The new Regulations will probably also make provision for an additional factor namely potential damage to "resource quality". In determining hazard potential rating the PLL and PEL are considered separately and the highest rating obtained, is accepted. The present structural condition of a dam does not influence its category classification. Only the size of the dam and its hazard potential with respect to downstream development are taken into account. The relation between category classification, size class and hazard potential is given in the table below.

Number of classes: 3

35.2.2. Dam classes overview

Size class	Hazard potential rating		
H = maximum wall height (m)	Low PLL=0 PEL: minimal	Significant PLL≤10 PEL: significant	High PLL>10 PEL: great
Small (5<H<12)	Category I	Category I	Category I
Medium (12≤H<30)	Category II	Category II	Category II
Large* (H≥30)	Category III	Category III	Category III

The size class "large" should not be confused with the International Commission on Large Dams (ICOLD) definition of a large dam. Information used for classification includes the position (distance downstream as well as distance from the watercourse and height above the river bed) of all development and infrastructure downstream of the dam, the size of bridge openings and the number of vehicles per day which use a particular bridge.

35.3. TECHNICAL FRAMEWORK

No norms and standards available, but guidelines on three topics have been made by "unidentified" authors (probably independent of DWA/DSO):

- Design or evaluation criteria

- Methods of Analysis

- Loads and Material assumptions

In addition, a guideline on floods is mentioned:

SANCOLD Guidelines on Floods (SANCOLD: 1991)

36.   SOUTH KOREA

Reporters: Shin, Dong-Hoon, Park, DongSoon

36.1.   MAIN PRINCIPLES OF DAM SAFETY MANAGEMENT

The primary principles of dam safety management in South Korea are for the enhanced level of dam safety technology and proper execution of inspection, monitoring, and preparedness. To ensure the quality of dam safety programs research and development of relevant technology are promoted and put into efficient use and management with the aim to contribute in the promotion of public welfare and the development of the national safe environment.

The inventory performed by the Korea National Committee on Large Dams (KNCOLD) showed a total of more than 18,000 dams in Korea, of which 1,265 are the large dams (over 15 m in height or having a storage more than 3 million m³) as of year 2010.

36.1.1. Legal framework for dam safety

There are primarily two national acts (including subsidiary enforcement decree, and enforcement rule) for directly enforcing dam safety in South Korea as follows:

- Act on Safety Management and Disaster Prevention of Reservoirs and Dams enacted in 2008 by the National Emergency Management Agency. This act directly concerns dam safety nationwide. The main purposes are to specify dam safety management, pre-check and maintenance to prevent disaster, and emergency action plan in case of dam failure.

- Special Act on the Safety Management of SOC Facilities established in 1995 by the Ministry of Land, Transport and Maritime Affairs. It specifies in detail the contents of safety control for existing facilities including large dams.

Besides the above two acts (including subsidiary enforcement decree and enforcement rule), there are other important acts that also affect dam safety management:

- The Act of Earthquake Disaster Measures (enacted as 2008)

- Construction Technology Management Law (enacted as 1987): deals with safety control for dam construction works

- Disaster and Safety management Act (enacted as 2004): covers comprehensive safety control for natural and man-made disaster etc.

The act on Safety Management and Disaster Prevention of Reservoirs and Dams can directly designate high-risk dams from the result of safety inspection and repairing project.

State/Country	Law/Act concerning dam safety	Decrees etc. concerning dam safety
South Korea	yes	Yes

36.1.2. Responsibilities for dam safety

The act on Safety Management and Disaster Prevention of Reservoirs and Dams designates the responsibility of dam owners for the dam safety such as keeping safety management standard, making an effort to ensure safety through regular inspection, in-depth safety inspection, remediation and reinforcement, and emergency action plan.

Dams in South Korea are owned and managed by several different organizations. The three major dam owners are K-water (Korea Water Resources Corporation) and KRC (Korea Rural Community Corporation), which are government invested companies, KHNP (Korea Hydro & Nuclear Power Co. Ltd.) which is a denationalized public company.

The Central Safety Management Committee, headed by the Prime Minister, supervises and coordinates the overall policy related to disaster and safety of dams. The Coordination Committee of the Central Safety Management Committee, under the Minister of Public Administration and Security, is in charge of the overall process of negotiations and coordination with regard to tasks delegated by the Central Committee.

Eight subcommittees headed by the ministers of various government departments help ensure the seamless operation of the Central Committee, and the Safety Management Committee of each city, county, district, and province is headed by the chairmen of the autonomous bodies in the local areas. In addition, the Central Disaster Safety Measures Headquarters (CDSCH), headed by the Minister of Public Administration and security, is committed to ensuring a swift and efficient response to large-scale disasters, and supports incident response units at the relevant government departments and offices. The CDSCH also operates in local autonomous jurisdictions such as cities, provinces, counties, and districts.

36.1.3. Arrangements for independent dam safety supervision

By the Special Act on the Safety Management of SOC Facilities Act, the KISTEC (Korea Infrastructure Safety and Technology Corporation) has been designated as a specialized inspection public unit and dam classification is specified for priority of inspection as follows. There are three categories of usual safety inspection: regular, principal, and urgent inspections by the act. The most detailed inspection is separately performed by KISTEC periodically according to the safety grade by the act.

Safety Grade	Principal Inspection	In-depth Safety Inspection
A	More than once/3 years	More than once/6 years
B, C	More than once/2 years	More than once/5 years
D, E	More than once/1 year	More than once/4 years

36.2. DAM CLASSIFICATION

36.2.1. General principles of classification

The criteria for classification are the dam purpose and reservoir capacity.

The safety grade of a dam has to be determined after dam inspection results.

The same criteria apply to all types of dams and reservoirs.

Number of classes: 2

Safety grade of dams: A – E

36.2.2. Dam classes overview

Under the Special Act on the Safety Management of Facilities, dam facility classification was divided by the 1st and 2nd class with significance and scale of facilities.

Dam facility Classification	State
1st class	Multi-purpose dams, power generation dams, flood control dams, and water supply dams over 10 million m³ of reservoir capacity About 74 dams are classified as 1st class
2nd class	Regional water supply dams over 1 million m³ of reservoir capacity 457 dams are classified as 2nd class

Dam safety grade is classified as one of the following categories after dam inspection, according to the Special Act on the Safety management of Facilities.

Safety grade of dams Classification	State
A (Very good)	No technical issues or problems
B (Good)	Found minor defects on subsidiary parts, but no impact on the performance. A little repairing may be necessary for durability enhancement
C (Medium)	Found minor defects on main parts or extensive defects on subsidiary parts, but no impact on the safety of overall structure. Repairing may be necessary to prevent the loss of durability and performance for main parts, or simple reinforcing may be needed on subsidiary parts
D (Not good)	Found defects on main parts so that urgent repairing and reinforcement are necessary. Need to decide whether the use of a dam should be restricted.
E (Severe)	Dangerous situation for the safety of structure due to significant defects on main parts. Immediate use restriction and reinforcing and rehabilitation are necessary

36.3. TECHNICAL FRAMEWORK OVERVIEW

There are no specific technical guidelines, however Construction Technology Management Law (enacted as 1987) deals with safety control for dam construction works, i.e. inspection and quality control of dam construction, safety management of field works, etc.

36.4. SUPPLEMENTARY INFORMATION

Dam safety is one of the utmost areas in South Korea for sustainable security of dams. Many up-to-date research projects about dam design, construction, monitoring, and rehabilitation have been performed including maintenance technologies, enhancing flood control, internal erosion, risk-based hazard analysis, aging effects, seismic resistance, etc.

Also, since social and environmental conditions regarding dam construction have been rapidly changing, the demand from local administrations and residents for participation in dam projects is on the rise. In order to accommodate these demands, the Korean government has decided to establish a new dam policy. For that reason, Korea is moving forward to construct medium and small size dams, which are environmentally responsible and in harmony with environment and development. A guideline for "environment friendly dam design and construction" to emphasize environmental feasibility aspects has been prepared. Also, Korea pursues aggressively to set forth environment-friendly dam construction plans to minimize environmental damages, to reduce unfavorable environmental impact and to prepare alternative schemes for disrupted environment.

37. SPAIN

Reporter: Juan Carlos De Cea

37.1. MAIN PRINCIPLES OF DAM SAFETY MANAGEMENT

Spain, officially the Kingdom of Spain, is composed of 17 autonomous communities and 2 autonomous cities, Ceuta and Melilla. The concession for the use of water is currently assigned to the River Basin Authorities (Spain is subdivided in 9).

There is legislation concerning dams and dam safety, consisting of:

- Ministry Orders

- Guidelines issued by Spanish National Committee on Large Dams (SPANCOLD)

37.1.1. Legal framework for dam safety

The use of water by means of dams is regulated by a Royal Decree (RD) updated in 2008 (January). The RD includes a Chapter that deals with the Dam Safety. The RD is applied only for dams higher than 5 m or dams with a volume of reservoir higher than 100,000 m³.

In addition to the Royal Decree updated in 2008, the following legislation is applied: *

- "Instruction for the Project, Construction and Operation of Large Dams", Ministry of Public Works, Ministry Order, 1967.

Synopsis:

It is a technical very detailed and rigid norm in some aspects that only applies to large dams. It was created, mainly, for projecting and constructing dams. Currently it is just applied to private dams.

- "Directriz Básica de Planificación de Protección Civil frente al Riesgo de Inundaciones" Ministry of Interior. Ministry Order, 1994,

Synopsis:

It is a standard of public safety that includes dam failure as a factor of risk on the downstream population. It obligates to classify all the dams into three categories (A, B or C (see below)), depending on the damages that could be caused downstream in the event of a potential failure or incorrect functioning. It includes the Planning of Emergencies in the event of a potential dam failure or incorrect functioning. All the dams classified in the categories A and B, must have implemented the corresponding Emergency Action Plan for an adequate operation.

- "Technical Regulation on Dams and Reservoirs Safety", Ministry of Environment, Ministry Order, 1996.

Synopsis:

It refers to all dams, large or small ones, and it is a framework that points out the criteria that should be kept in mind in each one of the stages of dam's life, but in an open form. It is addressed especially to dam operation and maintenance. The Technical Regulation defines clearly the holder's figure and its liabilities in all the phases of the dam's life. It defines, clearly, which are the commitments of the Dam Safety Organization. It forces to carry out periodic safety inspections by a specialized team different from the dam operation team.

With respect to the project, and more exactly for the design of the spillway and outlets, it includes furthermore the design flood, a new one: The Extreme Flood. Hydraulic safety acquires, as consequence, the greatest importance. In that same line, it establishes the impossibility to carry out the first filling for a new reservoir if it is not approved and has implemented the corresponding Emergency Action Plan.

Also, as a great novelty it establishes the need to create and revise periodically the Technical Archive of the dam, to which it refers in a reiterative form on great part of its articulate. In this sense, the main conclusion that one could extract is that a poorly documented dam could not be considered safe.

State/Country	Law/Act concerning dam safety	Decrees etc. concerning dam safety
Spain	yes	Yes

37.1.2. Responsibilities for dam safety

The owner has the legal responsibility for dam safety and any damage the dam can create in case of problems, incidents, failure or during its normal operation.

a) Owners

The Owner is in charge of the responsibility of dam safety. About 70% of the Spanish dams are owned by private owners. The remaining dams are owned by public entities.

Administrative organisation

The following main organisations are in charge of dams:

- Dam Authority (currently named "Seguridad de Infraestructuras", Infrastructure safety") checks those projects relevant to large dams; survey of the construction and first filling phases; supervision of the surveillance and control activities carried out by the owner during the operation, evaluation of the results of safety re-assessment for existing dams; evaluation and approval of rehabilitation or repair works; supervision of technical activities related to the preparation of emergency plans; participation to the updating of Regulation and technical standards.

- Autonomous Administrations: Approval of projects involving "small dams". Supervision of the activities relevant to the construction and operation of "small dams".

- The Civil Protection Authorities are in charge of the management of possible emergency situations and of the rescue of the population in case of incidents.

37.1.3. Arrangements for independent dam safety supervision

For the "large" dams" the Ministry of Environment is responsible of the technical evaluation and approval of new projects and supervise the actions of the owner for the safety of the dams in operation (see the section above).

The "small" dams" are controlled now, since January 2008, by the Autonomous Governments with its regional laws.

37.2. DAM CLASSIFICATION

Two types of classification are adopted in Spain. One is done according to the dam dimensions, and another according to the downstream potential risk in case of failure or malfunctioning.

The "large dams" subjected to the national Dam Authority are defined by the following dimensional parameters: height H > 15 m, or dams between 10 m and 15 m with a reservoir volume V > 1,000,000 m³. The dam height is the difference between the elevation of the crest and the lowest elevation of the foundation.

With respect to Risks Dam's Classification Criteria used in Spanish Regulation, are the following:

Dam Category	Risk for Population	Essential Services[1]	Material Damages	Environmental Damages
A	Serious effect on towns or more than 5 inhabited dwellings	Serious effect on Essential Services	Very Serious	Very Serious
B	Would affect a small number of dwellings (from 1 to 5)		Serious	Serious
C	Incidental loss of life (no inhabited dwellings in the area)		Moderate	

[1] Essential services are considered to be such which are indispensable for the performance of a population around 10,000 inhabitants (supplies and welfare, power supply, communications, transport, etc.).The responsibility for smaller dams is assigned to the Autonomous Governments (Spain is subdivided in 17 Autonomies).

37.3. TECHNICAL FRAMEWORK OVERVIEW

Besides the technical topics comprised in the Legislation, the following documents composes the dam safety framework.

There is a group of Guidelines published by the Spanish National Committee on Large Dams, that act as recommendations, without legal obligation, but its use is now practical unanimous by the dam community:

1	Dam Safety
2	Criteria for the design of dams and appurtenant structures Volume 1º: Concrete Dams Volume 2º: Embankment Dams
3	Geological and Geotechnical studies and Prospecting for Materials
4	Design Flood
5	Spillways and Outlets
6	Dam Construction and Quality Control
7	Monitoring of Dams and Foundations
8	Risk Analysis applied to management of dam safety

Spain is preparing in this moment three technical guidelines, as basic regulation, for the classification of dams depending on the potential risk downstream, for Emergency Action Plans and its implementation, for dam decommissioning, for the project, construction and operation, etc.

38.   SRI LANKA

Reporter: Badra Kamaladasa

38.1.   MAIN PRINCIPLES OF DAM SAFETY MANAGEMENT

Sri Lanka is an Island in the Indian Ocean which has a history over 2500 years. It is a unitary republic state with 25 administrative districts. Sri Lanka is ruled by an Executive President Elected democratically by the people of the country. Irrigated agriculture place a vital role in the economy of the country. Hence, dams and reservoirs are given priority in the state budget. There are nearly 350 medium and large dams in Sri Lanka. All the dams are owned and managed directly by the following four central Government institutions and provincial councils.

- Irrigation Department (ID)

- Mahaweli Authority (MASL)

- Ceylon Electricity Board (CEB)

- National Water Supply & Drainage Board (NWS & DB)

- Provincial councils (PC)(nine of them)

ID and MASL are under the Ministry of Irrigation and Water Resources Management, CEB comes under the purview of Power and energy and NWS & DB under the Ministry of Water Supply and Drainage.

In addition, there are more than 12,000 small dams that form village level reservoirs providing water to village level requirements. These schemes are managed by the Farmer Organizations under the direction of Provincial Councils and Agrarian services Departments.

38.1.1. Legal framework for dam safety

There are several national acts and subsequent regulations covering aspects of dam safety.

- State Land Ordinance No. 08 of 1947

- Irrigation Ordinance No. 32 of 1946

- Mahaweli Authority Act No. 23 of 1979

- Environment Act No. 48 of 1980

State/Country	Law/Act concerning dam safety	Decrees etc. concerning dam safety
Sri Lanka	yes	Yes

38.1.2. Responsibilities for dam safety

The owner of medium and large dams (ID, MASL, NWS & DB of CEB) is responsible for the dam safety, from the design phase to maintenance, and strictly liable for consequences of dam failure. Each dam owner has set up a specialized branch to look after the concerns of dam safety.

Dam Owner	Branch
ID	Dam Safety Branch
MASL	Head Works Division
CEB	Dam Safety, Environment and civil structures maintenance
NWS & DB	Production - Western
PC	Engineering Divisions

38.1.3. Arrangements for independent dam safety supervision

Supervision of dams is performed by regional heads of each institution. In ID all dams are kept under strict supervision of an Irrigation Engineer and an Engineering Assistant assigned for the job. During the rainy season and during the flood period both officials inspect the dam twice a week and the regional head (Regional Director) inspects weekly. In normal situations inspection is done in the following manner; Engineering Assistant – weekly, Irrigation Engineer – monthly and Regional Director – once in the season.

In MASL each major dam is under the purview of an Engineer in-charge (EIC) who is overseen by the Director, Head Works Division. Each EIC prepares a weekly diary for each dam to inspect and operation and maintenance.

In CEB, strict supervision of dams is performed by respective chief Engineer. Maintenance, periodic inspection and monitoring are done by a civil Engineer who reports to the Deputy General Manager of respective dams. Annual inspection is carried out by a team comprised of chief engineers in the civil, electrical and mechanical filed, who are responsible to the Deputy General Manager of Dam Safety, Environment and civil structures maintenance.

NWS&DB, which owns only two large dams, has no special arrangements for independent Dam supervision.

In nine Provincial councils, no specific dam safety approach is adopted, since they possess a limited number of medium and small dams. In two provincial councils where two to three large dams are situated, technical advice of Central irrigation Department is sought when a dam safety is issue is surfaced.

38.2.   DAM CLASSIFICATION

38.2.1. General principles of classification

Classification of dams in Sri Lanka has been done according to physical features.

Classification 1 – Type of Dam – rock fills dam, earthen dam, concrete dam etc.

Classification 2 – Height of the Dam and capacity of reservoir.

- Large Dams - dam height more than 15 m or dam height between 5 m and 15 m and compound volume is more than 3 mcm. There are 137 dams categorized as large dams.

- Medium Dams - dam height between 5–15 m and compound volume is less than 3 mcm or dam height less than 5 m and compound volume sufficient to serve irrigable area more than 80 ha.

- Small Dams - dam height less than 5m and compound volume is less than 3 mcm and irrigable area less than 80 ha. There are about 12,500 such small dams in the country.

38.3.   TECHNICAL FRAMEWORK OVERVIEW

The following dam inspection and supervision guidelines are given by the authorities to the regions.

Irrigation Department

- Department Circulars No. 42 of 1978 on "Attention on Irrigation Schemes During North East Monsoon".

- Dam Safety circular No. 15 of 1998 on "Quarterly and Monthly Inspection Of Head Works And Reporting Procedure"

- Dam Safety Instructions 1/2011

Mahaweli Authority of Sri Lanka

Follows instructions similar to the SEED Manual (USBR Guideline) adopted to the local conditions. Generally, the following procedures are adopted for dam surveillance and inspection:

- Weekly diaries are maintained by the engineer in charge

- In house supervision by team of engineers (annually)

- Independent local Experts Supervision (annually)

- International Experts Supervision (5 yearly)

CEB and NWS &DB

Since the number of dams under the purview of these institutions are quite limited, no special guideline is prepared for dam inspection and supervision procedures.

Provincial Councils

Engineering sub departments of the nine provincial councils adopt different types of procedures for dam surveillance which are embedded in annual operation and maintenance programs.

39.   SWEDEN

Reporter: Maria Bartsch

39.1.   MAIN PRINCIPLES OF DAM SAFETY MANAGEMENT

Sweden is a unitary state divided into 21 counties. Each county has a County Administrative Board, which among other tasks is the supervisory authority for dams in the county. On the national level the government has assigned Svenska Kraftnät (the Swedish National Grid) to promote dam safety and provide guidance on dam safety supervision to the county administrative boards.

There are more than 11.000 dams in Sweden, and about 200 of these are large hydropower or tailings dams higher than 15 meters.

39.1.1. Legal framework for dam safety

Sweden has no specific law or regulations concerning dam safety. Rather several different statutes are applicable to dams and dam safety, and requirements are expressed in all embracing and generally formulated rules for activities that may have consequences on human health and the environment. The most important requirements on dam safety are in the Environmental Code (which includes the former Water Act), the ordinance on operator's self-control and in the Civil Protection Act.

State/Country	Law/Act concerning dam safety	Decrees etc. concerning dam safety
Sweden	yes	No*
*The Ministry of Enterprise has initiated a state inquiry 2011–2012. The task is to draft a more specific dam safety legislation including for example compulsory consequence classification for dams and reporting routines from the owner to the supervisory authority. Also, the organization of dam supervision, supervisory guidance and promotion of dam safety is under review.		

39.1.2. Responsibilities for dam safety

The dam owners are responsible for dam safety, and strictly liable for consequences of dam failure.

The county administrative boards are the supervisory authorities for water activities, including dam safety, in each region. Supervisory guidance is provided by Svenska Kraftnät.

Arrangements for independent dam safety supervision.

Governmental supervision of dams is performed by the 21 County Administrative Boards. Svenska Kraftnät provides supervisory guidance to the County Administrative on routines for dam safety supervision of dams classified in consequence class 1A, 1B and 2, see below.

39.2. DAM CLASSIFICATION

39.2.1. General principles of classification

The classification system is not given in the law but in industry guidelines (RIDAS). However Svenska Kraftnät – the national dam safety authority - recommends the county administrative boards – the regulator - to ask all dam owners to classify their dams according the consequence classification system in RIDAS, and report the classification to them.

The consequence class of the dam is determined by the dam owner based on the most serious consequences of dam failure regarding the risk of:

* loss of human life or serious injuries,

* the loss of societal values (public installations),

* environmental values, and

* economic damage.

Number of classes: 4, denoted 1A, 1B, 2 and 3 where 1A is the highest class. Dams that may cause loss of life in the event of a failure fall into class 1A and 1B.

In 2012 the classification scheme in RIDAS has been revised in order to introduce a new highest class denoted "1+" for dams where a dam failure may cause extreme consequences (a national crisis). The new denotations of the classes are 1+, 1, 2 and 3.

39.3. TECHNICAL FRAMEWORK OVERVIEW

The following guidelines relevant to dam safety have been prepared in Sweden:

* Guidelines for design flood determination for dams (rev. 2007) issued by Svenska Kraftnät, SwedEnergy and SveMin (that is the national dam safety authority in cooperation with the industry organizations for the power industry and the mining industry).

* Dam safety guidelines by the industry organisations of the dam owners; RIDAS prepared for hydropower dams by SwedEnergy and GruvRIDAS for tailings dams by SveMin. Besides technical advice the guidelines contain advice on dam safety management, supervision, maintenance and emergency preparedness planning. The dam owners applying RIDAS use a consequence-based approach for their dam safety work, where the demands on dam safety and the extent of the dam safety activities are based on the potential consequences of dam failure.

RIDAS comprises guidance on:

* Consequence classification

* Organisation, competence and documentation

* Dam design and construction; design load determination, embankment dams, concrete dams, discharge facilities

* Operation, surveillance and monitoring, and maintenance

- Emergency preparedness, and

- Dam safety audits

39.4. SUPPLEMENTARY INFORMATION

More information on dam safety management, legislation and classification can be found on the webpage of Svenska Kraftnät, www.svk.se.

A description of the Swedish system for dam safety and ongoing development is given in the paper "Review of the Swedish system for dam safety", Bartsch, M. and Engström Meyer, A., Svenska Kraftnät, Q.93 ICOLD congress Kyoto 2012.

40. SWITZERLAND

Reporter: Marc Balissat

40.1. MAIN PRINCIPLES OF DAM SAFETY MANAGEMENT

Switzerland, in its full name the Swiss Confederation, is a federal republic consisting of 26 cantons, with Bern as the seat of the federal authorities. Citizens may challenge any law or introduce amendments to the federal constitution through referendums and initiatives which makes Switzerland a direct democracy.

The Federal Constitution adopted in 1848 is the legal foundation of the modern federal state. It is among the oldest constitutions in the world. A new Constitution adopted in 1999 divides the powers between the Confederation and the cantons and defines federal jurisdiction and authority. There are three main governing bodies on the federal level: the bicameral parliament – Federal Assembly (legislative), the Federal Council (executive) and the Federal Court (judicial).

There are over 200 large dams (second to the information on the official page of Swiss national committee) in Switzerland and 155 of them are higher than 15 m. The first dam was constructed in 1872, but the greater part of the dams have been constructed between 1950 and 1970 (over 80 dams higher than 15 meters), when Switzerland´s economic development increased also the demand for electrical energy. Most dams have been built for hydropower production. In the last decades the construction of dams has decreased significantly and activities have concentrated on maintenance and rehabilitation of the existing hydro schemes.

The large dams in Switzerland are under the supervision of the Federal Office for Energy, which is the federal supervisory authority. Dams that do not meet the size specifications are not explicitly subject to federal supervision and are under the supervision of the cantons.

40.1.1. Legal framework for dam safety

The main regulation that concerns dam safety is:

- Federal law on water regulation (RS 721.10 June 22, 1877, Bundesgesetz über die Wasserpolizei) cancelled in 2010 and replaced by the Federal law on water impounding works (RS 721.101 Oct. 1, 2010, Bundesgesetz über die Stauanlagen). The Federal law defines the safety measures for impounding works (construction and operation phases, as well as emergency case), the liability and the surveillance responsibility.

- Decree on water impounding works (RS 721.101.1 Oct. 17, 2012, Stauanlagen-Verordnung, StAV) contains the basic legal texts. The 1998 Decree redistributes the supervisory authority between the federal and cantonal authorities. It is applicable to all dams that are higher than 10 meters or dams with a height of at least 5 meters and a minimum storage capacity of 50,000 cubic meters.

Apart from the law and the decree mentioned above there are more federal laws and decrees that concern some specific aspects of reservoirs (e.g. use of hydraulic force, river development, water protection, fishing, land planning, etc.).

State/Country	Law/Act concerning dam safety	Decrees etc. concerning dam safety
Switzerland	yes	No*

40.1.2. Responsibilities for dam safety

The one entity responsible for dam safety is the dam owner. The owner is obliged to carry out the primary surveillance of the dam which includes the regular visual inspections, instrumentation monitoring, measurements, plausibility check, dry and wet tests of the mechanical equipment, such as bottom outlet valves, spillway gates, etc. The extent of surveillance is defined in the surveillance program which is mandatory by regulation. All observations and results of monitoring are summarized in a yearly report.

The surveillance and the collection of monitoring data are performed by the dam guardian. At each dam an experienced engineer carries out the yearly inspections, checks the data, analyses them and makes comments on the condition and the performance of the dam.

40.1.3. Arrangements for independent dam safety supervision

The Dam safety in Switzerland is organized on three pillars:

1. Structural safety - Assuring the structural safety (design criteria that strictly require no submergence even during the largest flood event or strongest earthquake and that take into consideration also the adequacy of drawdown capacity, the geological conditions and the resistance to seismic loadings).

2. Operation, monitoring and maintenance:

    –   Adequate operation, performance of monitoring and maintenance works (visual inspections of the dam, regular reading of monitoring equipment, check of hydro-mechanical equipment operation, repair of defective equipment) (level I)

    –   Supervision of monitoring and analysis. Control and analysis of performed monitoring, operation and maintenance by an experienced engineer (level II)

    –   Review and assessment performed by an independent expert every five years (level III).

3. Organization of early warning of population and civil protection (emergency preparedness plans). The mitigation of residual risk is subdivided in three elements:

    –   Alarm system and early warning (3 levels)

    –   Flood mapping

    –   Preparedness plan

For all dams higher than 40m with a reservoir capacity larger than 1 million m^3 a five-year comprehensive review is required that has to be prepared by a licensed dam expert accompanied by a geology expert.

For dams higher than 10 m and dams with a particular hazard the principles of dam safety are applied under the supervision of the Federal Office for Energy. Dams which are not included in these two categories are controlled by cantonal authorities.

The performance of dam surveillance and dam safety care is organized at several levels to reduce or minimize the risk of an undetected critical feature or potential hazard.

```
┌─────────────────────────────────┐
│   Federal (cantonal) authorities │
└─────────────────────────────────┘
                 │
                 ▼
┌─────────────────────────────────┐
│     Expert or team of experts    │
└─────────────────────────────────┘
                 │
                 ▼
┌─────────────────────────────────┐
│      Experienced engineer        │
└─────────────────────────────────┘
                 │
                 ▼
┌─────────────────────────────────┐
│   Dam owner – dam guardian       │
└─────────────────────────────────┘
```

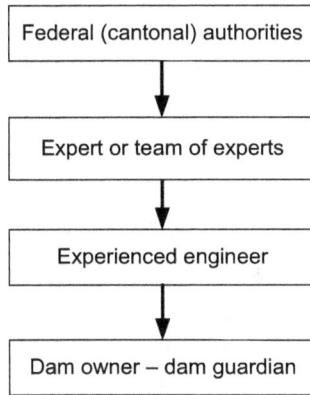

These multiple levels of surveillance and review (dam guardian, experienced engineer, expert(s), federal authorities) allow minimizing the risk of having an undetected critical feature or a potential hazard.

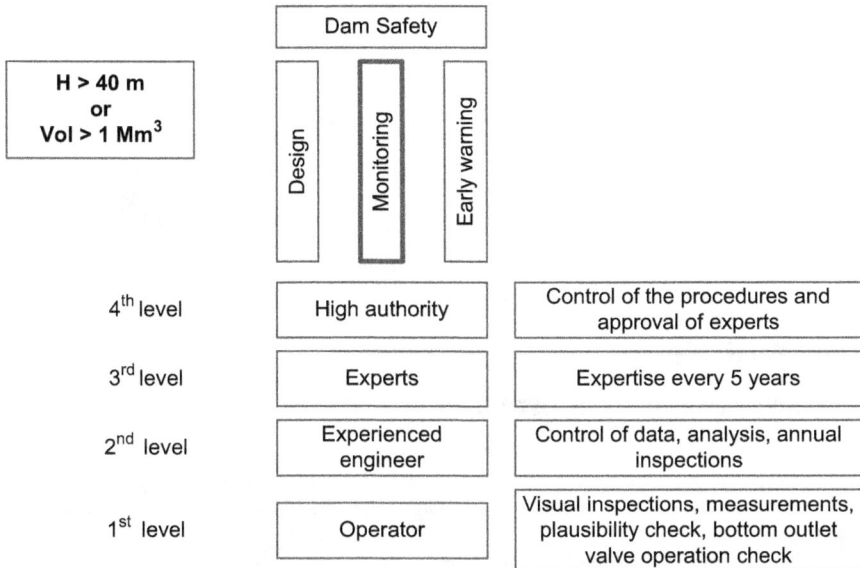

	Dam Safety	
H > 40 m **or** **Vol > 1 Mm^3**	Design / Monitoring / Early warning	
4th level	High authority	Control of the procedures and approval of experts
3rd level	Experts	Expertise every 5 years
2nd level	Experienced engineer	Control of data, analysis, annual inspections
1st level	Operator	Visual inspections, measurements, plausibility check, bottom outlet valve operation check

Fig7.1
Organization of supervision (Source: http://www.sesec.org/pdf/7/SESEC7_Droz.pdf)

40.2. DAM CLASSIFICATION

40.2.1. General principles of classification

The dam categorization follows the Federal Decree on the safety of water impounding works (1998) and the guidelines of the Federal Office of Energy (2003).

There is a basic division between dams controlled by Federal Office of Energy (dams higher than 10 m and dams with particular hazard) and other dams controlled by cantonal authorities. The categorization is only related to the destruction potential of the reservoir volume and dam height respectively to the size of the released water wave in case of a dam break. There is no quantified consideration of loss of lives or damage to property.

There are two classification criteria:

- classification by the hazard in case of uncontrolled release of water (based on potential velocity and depth of flood wave, combined with downstream housing, school or hospital building)

- classification in function of dam height and volume of impounded water for the structural analysis in case of earthquake (categories I, II and III)

40.2.2. Dam categories overview

Category	Description	Criteria
I	Dams that require a review and assessment by an acknowledged expert every 5 years	$H \geq 40m$ or $H \geq 10m$ and $V > 1'000'000m^3$
II	Dams that do not belong to Category I, but are still submitted to the direct federal supervision	$H \geq 25m$ or $H > 15m$ and $V \geq 50'000m^3$, or $H > 10m$ and $V \geq 100'000m^3$, or $H > 5m$ and $V > 500'000m^3$
III	All other dams that do not belong to Categories I or II	Exception: high hazard, defined in function of potential damage to people and property downstream of the dam

40.3.   TECHNICAL FRAMEWORK OVERVIEW

Technical framework that concerns dam safety - Guidelines published by the Federal Office for Energy (http://www.bfe.admin.ch/themen/00490/00491/00494/index.html?lang=en) are:

- Safety of water impounding works (ID: 926, 01.11.2002)

- Safety of water impounding works / Basis documentation for submission criteria (ID: 927, 01.06.2002 / Bischof R., Hauenstein W. et al)

- Safety of water impounding works / Basis documentation for structural safety (ID: 929, 01.12.2002 / Pougatsch H. et al.)

- Safety of water impounding works / Basis documentation for the safety in case of flood (ID: 3729, 01.06.2008)

- Safety of water impounding works / Basis documentation for monitoring and maintenance (ID: 928, 01.12.2002 / Pougatsch H. et al)

- Safety of water impounding works / Basis documentation for the safety verification in case of earthquake (ID: 930, 01.03.2003 / Darbre G. et al)

- Safety of water impounding works / Guideline for the verification in case of earthquake / Examples of dams with small height (ID: 931, 01.03.2003 / Darbre G. et al)

- Accelerogram samples to be applied according to dam foundation conditions (ID: 3773 to 3776, 01.07.2008 / Studer Engineering)

40.4. SUPPLEMENTARY INFORMATION

More information about dam safety management, legislation and classification can be found in:

[1] http://www.swissdams.ch/default_e.asp

[2] http://www.bfe.admin.ch/themen/00490/00491/00494/index.html?lang=en)

[3] http://www.sesec.org/pdf/7/SESEC7_Droz.pdf

[4] http://www.waterpowermagazine.com/Table.asp?sc=2040340&seq=2

[5] Mouvet, L. Müller, R.W., Pougatsch, H. – Structural safety of dams, according to the new Swiss legislation, Dams in a European Context: Proceedings of the ICOLD European Symposium, 25 to 27 June 2001, 2001, Taylor & Francis, p.p.271–286

41. TURKEY

Reporter: Tuncer Dincergok

41.1. MAIN PRINCIPLES OF DAM SAFETY MANAGEMENT

Turkey is a democratic, secular, unitary, constitutional republic with 81 administrative provinces.

There are around 1,630 dams registered [Under Construction (UC) + In Operation (O)] in Turkey and about 870 [692 (O) + 178 (UC)] of these are higher than 15 m and classified as large dams according to the ICOLD criteria. The majority of these dams are designed and constructed by DSI, but through Law No. 4628 (Electricity Market Law, dated 2001) the private sector is also authorized to build and operate Hydropower Power Plants for 49 years. At the end of this period, operational rights and ownership of these facilities shall be turned over to the Government as well, and they are growing in number.

41.1.1. Legal framework for dam safety

Turkey has no specific legislation for dam safety yet but the general requirements are identified in the following laws and regulations:

- Law Concerning the Organization and Duties of the General Directorate of State Hydraulic Works (DSI) – (1953)

- Civil Defence Act (CDA) – (1958)

- DSI Regulation on Protection against Flooding - (1982)

- The Environmental Law – (1983)

- Regulation on the Environmental Impact Assessment (EIA) – (2003)

- Regulation on Supervision Services for Water Facilities – (2011)

Country	Law/Act concerning dam safety	Decrees etc. concerning dam safety
Turkey	yes	Yes

41.1.2. Responsibilities for dam safety

The governmental organization, General Directorate of State Hydraulic Works (DSI), is the main authority in charge of dam safety studies as well as defining dam design measures and standards in Turkey. The current dam safety assessment procedures are realized by the coordination of Dam Safety Branch structured under the Dams and HEPPs Department and through the activities of relevant departments located within the DSI General Directorate (Dams & HEPPs, Planning & Investigation, Operation & Maintenance, and Geotechnical Investigations) and DSI Regional Directorates.

The dam owners are responsible for dam safety during all phases of the lifecycle of their dams from design phase to operation in principle. Most majority of the dams are owned by the Government but, as mentioned above, dam owners can be private (persons, companies, auto-producers, auto-producer groups, municipalities, etc.) as well.

DSI Dam Safety Branch is established within body of the Dams and Hydroelectric Power Plants (HEPPs) Department at the General Directorate of State Hydraulic Works (DSI) for the organization and coordination of the activities to be carried out by the public authority in case of a failure of a dam in operation that is built by the public or the private sector in order to:

- Prevent and mitigate the potential risks caused by both natural and technical reasons,

- Minimize the damages and casualties to the downstream, and

- Recover the related damages

41.1.3. *Arrangements for independent dam safety supervision*

Government owned dams are inspected regularly and "Inspection Reports" as well as "General Dam Safety Reports" are prepared and/or updated yearly by the DSI staff on site. These reports are checked and approved by the relevant departments of the General Directorate of State Hydraulic Works (DSI).

Obligation for the dams built and operated by the private sector in terms of reporting to a Governmental Organization or to an Authority concerning Dam safety (after the facility is put into operation) is under preparation and shall be valid in 2013. Still, these facilities are also being monitored by the private sector by carrying out measurement programs.

On the other hand, during temporary and final acceptance formalities carried out by DSI (General Directorate of State Hydraulic Works) related measurements are interpreted from the dam safety point of view and additional precautions are taken if deemed necessary for the dams built and operated by the private sector.

"Draft Guideline on Dam Safety Management" is released on October 12, 2012 at the "1st Congress on Dams" and shall be finalized within 1-year time in accordance with the received comments from the related parties.

Last but not least, according to the "Regulation on Supervision Services for Water Facilities (2011)" to be re-arranged soon and enacted in 2013, the supervision services for the dams built by the private sector shall be carried out by independent engineering and consultancy companies to be identified by the General Directorate of State Hydraulic Works (DSI).

41.2. DAM CLASSIFICATION

Dams in Turkey are classified as "Small Dams" and "Large Dams" according to the following criteria issued by the Circular letter of the General Directorate of State Hydraulic Works (DSI):

- If the height of dam (from the thalweg) $\leq$ 35 m, Dam Body Volume $\leq$ 600,000 m³ and Reservoir Volume $\leq$ 4 hm³ then it is called a "Small Dam"; the rest are categorized as "Large Dams".

There are single purpose and multi-purpose dams in Turkey. These can have either one or a combination of the following purposes:

- Electricity Generation, Irrigation, Drinking &Industrial Water Supply, Flood Control. (These are "normally water-filled" reservoirs).

Also, there are some Tailings Dams (filled with deposit material of a mine, etc.) or and Flood Protection Levees (Dry)

Dams in Turkey are classified according to a Factor of Risk that depends on:

- Reservoir Capacity,
- Height of Dam,
- Number of Inhabitants to be evacuated at Downstream,
- Potential Loss at Downstream,

To each of these factors, a score is attributed in accordance with the following table:

Factor of Risk	The Highest	High	Moderate	Low
Reservoir Capacity (A)	> 120 hm³ (6)	120– 1 hm³ (4)	1 – 0.1 hm³ (2)	< 0.1 hm³ (0)
Height of Dam (B)	> 45 m (6)	45 - 30 m (4)	30 - 15 m (2)	< 15 m (0)
Number of Inhabitants to be evacuated at downstream (C)	> 1,000 (12)	1,000 - 100 (8)	99 - 1 (4)	0 (0)
Potential Loss at Downstream (*) (D)	High (12)	Moderate (8)	Low (4)	None (0)

(*) Losses of Industrial, Agricultural, Natural Resources & Infrastructure Facilities

In case of Dam Failure:

TOTAL FACTOR OF RISK = A+ B + C + D	
A	Factor of Risk for Reservoir Capacity
B	Factor of Risk for Dam Height
C	Factor of Risk for the Number of Inhabitants Downstream
D	Factor of Risk for the Potential Loss Downstream

TOTAL FACTOR OF RISK	RISK CLASS	COMMENTARY
(0 - 6)	I	Low Risk Group
(7 - 18)	II	Moderate Risk Group
(19 - 30)	III	High Risk Group
(31 - 36)	IV	The Highest Risk Group

41.3. TECHNICAL FRAMEWORK OVERVIEW

General Directorate of State Hydraulic Works (DSI) is the main authority in charge of dam safety studies as well as defining dam design measures and standards in Turkey.

Besides using its own standards such as "Planning and Design Criteria for Large and Small Dams (1995)" and design directives such as "Directive on Spillway Design Flood Determination (2006)" and TS 500; DSI applies the generally accepted dam engineering design standards such as the procedures, codes and standards of US Bureau of Reclamation, US Army Corps of Engineers as well as ICOLD recommendations in addition to ACI and DIN standards. Additionally, some of the design measures are given in the Civil Works Technical Specifications, Geotechnical Investigations Specifications and etc. prepared by DSI.

Recently, the following Draft Technical Guidelines are released on October 12, 2012 at the "1stCongress on Dams" and shall be finalized within 1 year time in accordance with the received comments from the related parties including designers, consultants, engineering companies, dam owners, universities, and others:

- Guidelines on Dam Safety Management

- Guidelines on Design of Embankment Dams

- Guidelines on Design of Concrete Dams

- Guidelines on Design of Appurtenant Structures

- Guidelines on Seismic Design Criteria for Dams

- Guidelines on Hydrology for Dams

- Guidelines on Hydraulic Design of Dams

- Guidelines on Design of Electromechanical & Hydromechanics' Equipment for Dams

41.4. SUPPLEMENTARY INFORMATION

More information regarding dam safety management, legislation and classification can be found in:

[1] Dincergok, T., "Development of Dam Safety Concept in Turkey", Paper presented at the 75th ICOLD Annual Meeting International Symposium, St. Petersburg, Russia - June 27, 2007.

[2] Dincergok, T., "Dam Safety & Dam Security in Turkey", 2nd International Week on Risk Analysis as Applied to Dam Safety and Dam Security, Valencia, Spain – February 26, 2008

[3] "Guideline on Dam Safety Management"(Draft, issuedin Turkish), 1st Congress on Dams, Ankara, Turkey, 11–12 October 2012,

42.   UKRAINE

Reporter: Alexander Karamushka

42.1.   MAIN PRINCIPLES OF DAM SAFETY MANAGEMENT

Ukraine is a unitary state with 24 provinces and one autonomous republic.

There are about 29 dams in Ukraine which are large dams according to the ICOLD criteria. All of them are State owned dams.

42.1.1. Legal Framework for Dam Safety

The safety of hydraulic structures is specifically governed by:

* the Law "On objects of increased danger"(from 18.01.2001 № 2245-III)

* the Regulations "On the uniform state system of emergency prevention and response"

The law "On safety of hydraulic structures" and its by-laws are in the development stage.

State	Law/Act concerning dam safety	Decrees etc. concerning dam safety
Ukraine	yes	Yes

42.1.2. Responsibilities for Dam Safety

The dam owners are responsible for dam safety and are governed by regulatory acts of Ukraine.

42.1.3. Arrangements for Independent Dam Safety Supervision

State supervision and control in the fields connected with objects of increased danger are exercised by an authorized government body and its respective territorial agencies.

42.2.   DAM CLASSIFICATION

42.2.1. General Principles of Classification

Dams as well as other hydraulic structures are classified according to consequences of dam failure, with stricter safety demands for dams with large failure consequences. Classification is carried out in accordance with the requirements of the "State building code of Ukraine.

Hydraulic, power and melioration systems and structures, mine workings. Hydraulic structures. Fundamentals. SBC B.2.4-3:2010".

42.2.2. Dam Classes Overview

Dam classification is made based on the probability of loss of health and life, economic losses, loss of cultural heritage and suspension of communications, traffic, energy and functions of other networks. See following table:

Consequence class	Basis for classification						Failure probability
	Loss of health and life probability, number of people.			Economic loss probability (minimum wages)	Loss of cultural heritage, Class of objects	Suspension of communications, traffic, energy and other networks functioning, level	
	Residing in the object	Periodically being at the object	Being outside the object				
CC3	More than 300	More than 1000	More than 50000	More than 150000	National significance	State	$5*10^{-5}$
CC2-1	from 20 to 300	from 50 to 1000	from 100 to 50000	From 2000 to 150000	Local significance	Regional, local	$5*10^{-4}$
CC2-2	from 20 to 300	from 50 to 1000	from 100 to 50000	From 2000 to 150000	Local significance	Regional, local	$3*10^{-3}$
CC-1	Less than 20	Less than 50	Less than 100	Less than 2000			$6*10^{-3}$

42.3. TECHNICAL FRAMEWORK OVERVIEW

In their activities dam owners follow Exploitation guidelines for hydraulic structures, written for them by developers.

42.4. SUPPLEMENTARY INFORMATION

At the moment the law "On safety of hydraulic structures" and its by-laws are being reviewed and elaborated. The requirements on safety are developed considering the specifics of hydraulic structures at all stages of their life activity (from design to taking out of service), as well as to the state supervision procedures for dam safety, the functions of involved organizations in the safety sector, etc. Adoption of this Law (approximately in 2012) would require substantial structural changes in the organization of hydraulic structure safety.

Reporters: Bruce C. Muller / Brian D. Becker

43.1. MAIN PRINCIPLES OF DAM SAFETY MANAGEMENT

The United States of America (U.S.A.) is a federal republic consisting of 50 states with their own governments and laws/acts and orders/decrees.

43.1.1. Legal Framework for Dam Safety

Dam safety regulation is the responsibility of:

- federal government for dams owned by institutions of the federation, and

- individual state governments for all remaining dams (i. e. the large majority of dams in U.S.A.)

There are several dam safety laws and regulations with national validity. Publishers are different institutions of the federal governmental administration; such as the following institutions. Those that own dams are self-regulating:

- Federal Emergency Management Agency

- Federal Energy Regulatory Commission

- Natural Resources Conservation Service

- US Army Corps of Engineers

- US Bureau of Reclamation

Currently 49 of the 50 federal states have laws and further regulations provided for regulation of dams.

The following table gives an overview of dam safety legislation in U.S.A.:

State/ Federal State / Province / Territory	Law/Act concerning dam safety	Decrees etc. concerning dam safety
U. S. A. (Federation)	yes/d*)	yes
50 States	yes	yes
	(Alaska, Arizona, Arkansas, California, Colorado, Connecticut, Delaware, Florida, Georgia, Hawaii, Idaho, Illinois, Indiana, Iowa, Kansas, Kentucky, Louisiana, Maine, Maryland, Massachusetts, Michigan, Minnesota, Mississippi, Missouri, Montana, Nebraska, Nevada, New Hampshire, New Jersey, New Mexico, New York, North Carolina, North Dakota, Ohio, Oklahoma, Oregon, Pennsylvania, Rhode Island, South Carolina, South Dakota, Tennessee, Texas, Utah, Vermont, Virginia, Washington, West Virginia, Wisconsin, Wyoming)	(Alaska, Arizona, Arkansas, California, Colorado, Florida, Hawaii, Idaho, Illinois, Indiana, Iowa, Kansas, Kentucky, Louisiana, Maine, Maryland, Massachusetts, Minnesota, Mississippi, Missouri, Montana, Nebraska, Nevada, New Hampshire, New Jersey, New Mexico, New York, North Carolina, North Dakota, Ohio, Oklahoma, Pennsylvania, Rhode Island, South Carolina, South Dakota, Tennessee, Texas, Utah, Virginia, Washington, West Virginia, Wisconsin, Wyoming)
	no	no
	(Alabama)	(Alabama, Connecticut, Delaware, Georgia, Michigan, Oregon, Vermont)
d = direct demands regarding dam safety i = indirect demands regarding dam safety (for instance by requiring accordance with state of the art or by referring to special decrees etc.) *) Federal Dam Safety Laws:		

Agency	Laws	Date
Federal Emergency Management Agency	Dam Safety Act of 2006 (Public Law 109–460)	2006
Federal Energy Regulatory Commission	Federal Power Act of 1935	1935
Natural Resources Conservation Service	Watershed Protection and Flood Prevention Act (Public Law 83–566)	1954
US Army Corps of Engineers	National Dam Inspection Act	1972
US Bureau of Reclamation	Reclamation Safety of Dams Act	1978

43.1.2. Responsibilities for Dam Safety

The dam owners are responsible for safety of their dams in principle. Dam owners can be private (persons, companies), municipalities, states, or federal agencies.

43.1.3. Arrangements for Independent Dam Safety Supervision

Inspection schedules and responsibilities vary from federal state to federal state. 18 states use an owner-responsible inspection model.

Federal agencies that own dams are mandated to have dam safety programs by Presidential Directive to implement the Federal Guidelines for Dam Safety.

To coordinate the diverse arrangements of dam safety regulation in the United States, the National Dam Safety Review Board was created by the Congress. The board is not an enforcement authority, but it facilitates the duties to be done by all dam owners or dam operators. It supports the development of policy, training, and research for the benefit of the dam owners.

43.2. DAM CLASSIFICATION

43.2.1. General Principles of Classification

Both States and the Federal Government generally classify dams according to hazard potential, size, and risk.

Within 46 federal states (all but Alabama, Delaware, Florida and Vermont) classification criteria are established. Risk based tools are applied in 19 states.

Within the Federal Government, agencies have adopted risk-based categorization to varying degrees. Among them the US Bureau of Reclamation has fully implemented a system of assessing risk that categorizes dams into one of the following three categories: A - increasing justification to reduce or better understand risks; B - decreasing justification to reduce or better understand risks; C – evaluate risks thoroughly, ensuring "As-Low-As-Reasonably-Practicable" (ALARP) considerations are addressed. The US Army Corps of Engineers is in the process of implementing a similar system.

The number of classification groups and the detailed criteria vary by Federal Agency and State.

Classification, depending on kinds of reservoirs, varies by Federal Agency and State.

43.2.2. Dam Classes Overview

The general scheme for subdivision of dams into different categories can be generalized in the below following table. The scheme varies by Federal Agency and State, but most have three to four (typically three) classes for dam size and hazard potential.

Size classification		Hazard potential classification	
Class	Dimension of criteria	Class	Seriousness of criteria
1	small	1	low
2	intermediate (medium)	2	significant
3	large	3	high
(4)	(very large)	(4)	(very high)

With respect to the hazard potential, two specific design criteria are of most importance: Design Storm (Design Flood) and Design Earthquake. Both specific criteria vary from State to State.

Their typical mean variations, depending on the size of hazard potential, can be summarized as follows.

Hazard Potential	High	Significant/medium	Low
Design Storm or Flood	0,5 ... 1,0 PMF or 0,4 ... 1,0 PMP	0,25 ... 0,5 PMF or 0,3 ... 0,5 PMP	0,25 PMF or >/= 100-year-flood
Design Earthquake	MCE or 2500 ... 5000-year-earthquake	0,5 MCE or 1000 ... 2500-year-earthquake	< 0,5 MCE or 500 ... 1000-year-earthquake
PMF = Probable Maximum Flood; PMP = Probable Maximum Precipitation; MCE = Maximum Credible Earthquake			

US Bureau of Reclamation utilizes a probabilistic approach to seismic, hydrologic and static loadings to evaluate structural performance in a risk context.

4.3. TECHNICAL FRAMEWORK OVERVIEW

There are central national Dam Safety Guidelines published by the Federal Emergency Management Agency (FEMA). Different guidance documents introduced by the states also exist. Most of the sates (40 of the 50) have established design criteria for dams especially with regard to design flood (storms) and earthquakes.

Federal Guidelines for Dam Safety are:

- Emergency Action Planning for Dam Owners (FEMA 64) (1998)

- Hazard Potential Classification Systems for Dams (FEMA 333) (1998)

- Selecting and Accommodating Inflow Design Floods for Dams (FEMA 94) (1998)

- Model State Dam Safety Program (FEMA 316) (2006)

4.4. SUPPLEMENTARY REFERENCES

More information about dam safety management, legislation and classification can be found in the following documents:

[1] FEDERAL GUIDELINES FOR DAM SAFETY (FEMA 93), (2004)

[2] Federal Guidelines for Dam Safety: Earthquake Analyses and Design of Dams (FEMA 65)

[3] Federal Guidelines for Dam Safety: Glossary of Terms (FEMA 148)

44.   VIETNAM

Reporters: Pham Hong Giang & Dinh Sy Quat

44.1.   MAIN PRINCIPLES OF DAM SAFETY MANAGEMENT

Vietnam comprises 62 provinces. On the national level the Central Government has assigned Ministry of Agriculture and Rural Development (MARD) and Ministry of Industry and Trade (MOIT) to manage and assure safety of all registered dams, and provide guidance, recommendation on safety of small dams (with dam height under 15 m and/or storage capacity less than 3 million cubic meters) built and operated by local authorities or communities. Provincial Governments are assigned to cooperate with MARD and MOIT in promoting dam safety. Each province has a Department of Agriculture and Rural Development (DARD), which among other task is supervisory authority for dams in the province.

There are more than 800 registered large dams in Vietnam, all are higher than 15 meters and about 50 of these are hydropower dams.

44.1.1. Legal framework for dam safety

In aspect of dam safety there are several national Vietnamese legal documents, among which the paramount requirements are given in the Decree 72 December 2007 issued by Vietnam Government and Circular 34/2010 by MOIT.

Country	Law/Act concerning dam safety	Decrees etc. concerning dam safety
Vietnam	No	yes

44.1.2. Responsibilities for dam safety

The dam owners are responsible for dam safety, from design, construction to operation phase, and strictly liable for consequences of dam failure. Dam owner can be private (communities, companies), provinces or nation.

The dam safety authorities are MARD and MOIT for registered dams, and Provincial Authorities for small dams sited in the provinces.

44.1.3. Arrangements for independent dam safety supervision

Governmental safety supervision of dams in Vietnam is performed by Water Resources Directorate (WRD) under MARD and MOIT in cooperation with concerned DARDs. WRD provides supervisory guidance to DARDs on routines for dam safety supervision of dams.

44.2. DAM CLASSIFICATION

44.2.1. General principles of classification

The classification system is not given in the law but in regulation (QCVN 04–05 : 2012). According to the regulation class of dams is decided and issued by the Central Government, represented by MARD and MOIT.

The class of a dam is determined based on:

- Sizes of the reservoir created by the dam,

- Types and height of dam, and

- Geological condition at the dam site.

Number of classes: 4, denoted I, II, III and IV where I is the highest class.

Dams with dam height above 100 m fall into class I.

However, introduction of classification of dams based on consequences of dam failure is under way. According to that, the class of dams is determined based on:

- Potential for loss of life,

- Potential for property damage, and

- Adverse impact on environment.

44.3. TECHNICAL FRAMEWORK OVERVIEW

The following documents concerning dam safety have been prepared in Vietnam:

- Ordinance for operation and protection for hydraulic works issued by Vietnam Government

- Circular 34/2010 for safety management of hydropower dams issued by MOIT

- Decree 72/2007 for dam safety management issued by Vietnam Government; the Decree applies to all dams under MARD and hydropower dams under MOIT, and contains articles on:

  - Consequence classification

  - Dam design and construction

  - Safety management of dam design

  - Dam construction supervision

  - Documentation

  - Reservoir operating procedures

  - Operation of gates in spillways

- Instrumentation

- Operation and maintenance

- Inspection program

- Dam safety report

- Emergency preparedness plan

44.4. SUPPLEMENTARY INFORMATION

More information on dam safety management, legislation and classification can be found on the webpage of MARD www.agroviet.gov.vn and of MOIT www.moit.gov.vn.

A Dam Safety Manual prepared by MARD is to be released in December 2012 that contains update guidelines and recommendations on selecting and accommodating inflow design floods, earthquake analyses, and safety inspection of dams.

For Product Safety Concerns and Information please contact our EU
representative GPSR@taylorandfrancis.com
Taylor & Francis Verlag GmbH, Kaufingerstraße 24, 80331 München, Germany

www.ingramcontent.com/pod-product-compliance
Lightning Source LLC
Chambersburg PA
CBHW060306220326
41598CB00027B/4246

9 781032 456058